Morphological Studies of English Canals

University of Hull Publications

Occasional Papers in Geography No. 20

General Editor: H. R. Wilkinson

Professor of Geography in the University of Hull

MORPHOLOGICAL STUDIES OF ENGLISH CANALS

J. H. FARRINGTON

University of Hull Publications
1972

Frontispiece A mile north of Shipton-on-Cherwell, the Oxford Canal (right) joins the Cherwell for three-quarters of a mile. The tail of a canal lock may be seen, extreme right. The graceful iron bridge carrying the canal towing path over the river is one of a number on the canal. Looking north-west (upstream).

SBN 90048027-0

Printed in England at the
Lowgate Press by
Hull Printers Limited, Willerby, Hull HU10 6DH

Acknowledgements

The consent of the Archivist of British Transport Historical Records, London, to the use of BTHR materials for publication is acknowledged. Mr. E. H. Arnold, Mr. A. Blenkharn and Mr. K. Ramsden of British Waterways gave assistance in the course of the original research. Acknowledgement is made to Aerofilms Ltd. for permission to use an oblique showing Caen Hill locks (plate 6). Thanks are also due to the technical staff of the Department of Geography, University of Hull, who drew many of the maps for publication.

I am grateful to Professor H. R. Wilkinson for help in the preparation of this paper, and to Dr. J. H. Appleton for his valuable advice and criticism.

J. H. F. January, 1973

CONTENTS

List of Figures

List of Tables

List of Plates

Introduction

The fundamental part played by transport improvements in the Industrial Revolution and in agricultural and social improvement in the 18th and 19th centuries is widely acknowledged. It is often underlined as being an integral part of the general economic expansion of this period. But the study of the lines of transport themselves has not been given the attention it deserves, particularly in the case of canal transport (Appleton, 1965). And yet this is an important aspect of the transport revolution, since in many areas industry and population located along the transport lines; it was not simply a case of canal seeking industry and population, though of course these factors exerted a profound influence on the routes of canals. Furthermore, canals which passed through rural areas were the means of improving agriculture by carrying lime, manure and agricultural produce.

Therefore an understanding of the factors which governed the location of these transport lines assists in understanding the changes in population distribution and industrial location, and the growth of new industrial and urban areas which was characteristic of the 18th and 19th centuries in England. An appreciation of the value judgements which were made in the process of canal route selection assists in understanding the phenomenon of expansion in this period, since the growth of a canal network was itself part of that expansion.

The English canal system may also be seen as an important element in the sequence of evolving transport systems, being preceded by river and road, replacing both to a large extent for long-distance movement of goods, and itself being succeeded by rail and road transport.

The research on which this paper is based was carried out from 1966 to 1969 for a Ph.D. Thesis in the Department of Geography at the University of Hull. Eleven of the trunk canals of England were selected for study, and it is from this sample that examples will be taken (Fig. 1 and Appendix 1). The aim of the research was to examine the relationships which exist between canal routes and canal morphology and their environment, comprising factors influencing routing and morphology. These include factors which operated during the promotion and construction of the canals and

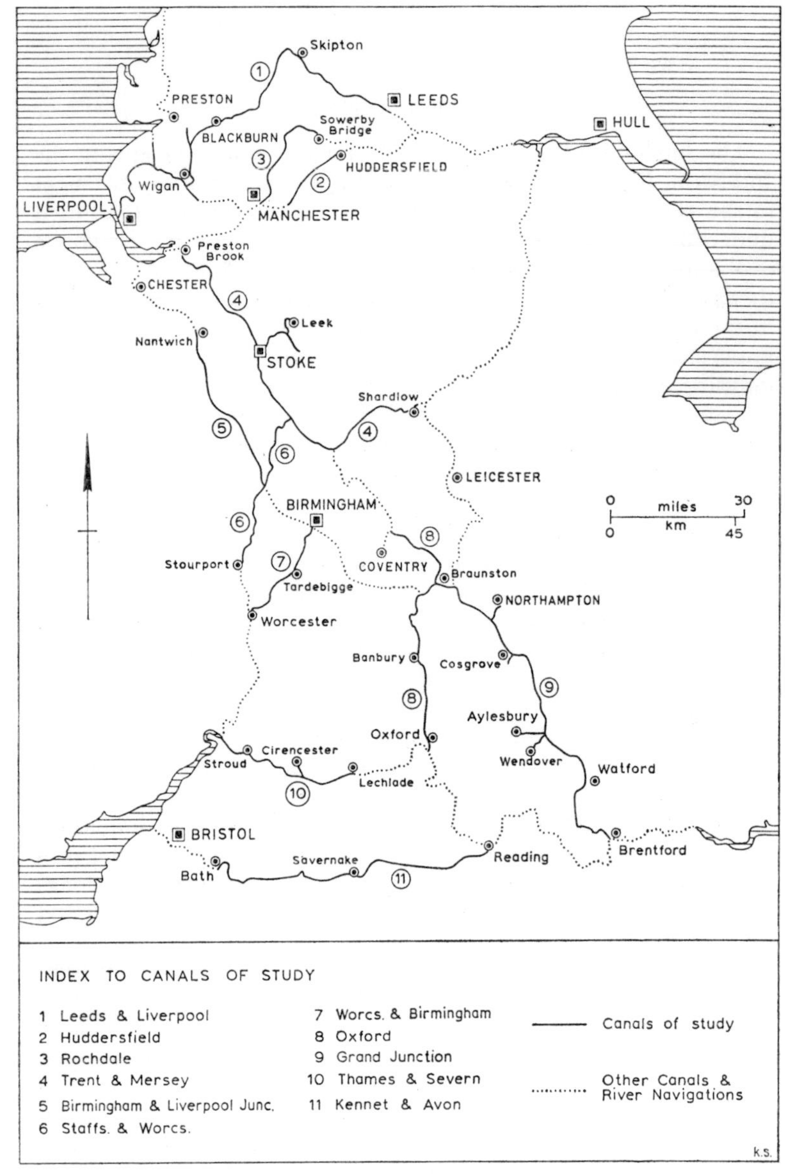

FIG. 1. THE ELEVEN CANALS

which may be classified as economic, engineering, financial, hydrological and topographic. Also included are less tangible influences such as inter-company rivalry and co-operation, opposition from vested interests, and the impact of personalities.

The procedure adopted in this paper is to specify and exemplify the findings made as a result of the examination of the interaction of these factors.

There are two ways of approaching the morphology of canals: (1) from the historical point of view, using contemporary sources to identify influences such as engineering capabilities, availability of finance, economic function of the canal, and interaction between different canal companies and vested interests; and (2) from the physical point of view, using field work and morphological indices to analyse relationships between canal and landform. Both approaches focus on canal morphology, and each contributes to an understanding of the morphology.

For reasons of space, the procedure here will be to concentrate attention on the second approach, introducing contemporary factors where relevant. For reasons of clarity the enquiry will be concerned with particular aspects of the environment-morphology relationship, but it should be understood that in practice there was a simultaneous interaction between the factors discussed. This emerges in the discussion of the examples given to support the findings.

CHAPTER I

The Crossing of Watersheds: Summit Level Morphology

THE NETWORK MODEL

MOST of the English trunk canals must cross major watersheds to connect the main terminal areas concerned. This is clearly seen if the network model of the main English canals is considered (Fig. 2). It is usual to visualise the trunk lines of the system as forming a cross connecting the four main estuaries and centring in the Birmingham area. More accurately this can be extended to a figure-of-eight by including the trans-Pennine and Thames-Severn routes. This gives six vertices; all were duplicated, and the trans-Pennine and Thames-Severn routes were triplicated, during the canal era. Thus the primary cross was made up of the following canals; the Trent & Mersey which made up two vertices in effect from the Mersey to the Midlands and from the Midlands to the Trent and Humber; the Staffordshire & Worcestershire Canal, connecting the Severn with the Midlands; and the Oxford Canal and the Coventry Canal, between them connecting the Thames with the Midlands. All four canal Acts were passed between 1766 and 1769; in those four years the outline of the canal system was laid out.

Two points should be made about this model. Firstly, the canals were opened for traffic at widely differing dates—the Staffordshire & Worcestershire as early as 1771, the Coventry not until 1790. Therefore the functional pattern of the cross was not established until some time after its conception—which is inevitable. Secondly the primary cross itself did not communicate with Birmingham and the heart of the Midland Plateau. This link was made by a number of canals, two of the most important being the Birmingham, and the Wyrley & Essington Canals, en-Acted in 1768, and 1792, respectively.

The two early east-west vertices were the Leeds & Liverpool Canal and the Thames & Severn Canal, authorised respectively in 1770 and 1783.

The secondary cross began with the Worcester & Birmingham Canal, en-Acted in 1791 to compete with the Staffordshire & Worcestershire Canal/Birmingham Canal route and with the Staffordshire & Worcestershire/Stourbridge/Dudley Canal route

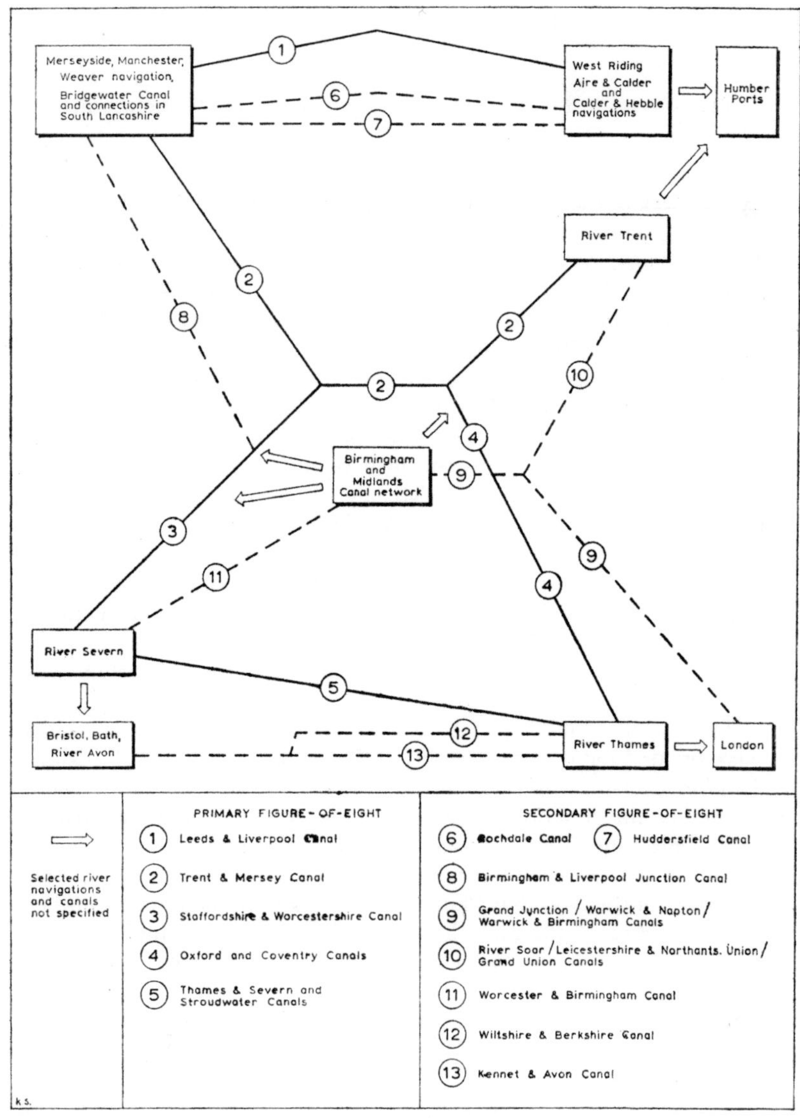

FIG. 2. 'THE CROSS' CANAL NETWORK

from the Severn to the Midlands. The Thames–Midlands vertex was duplicated by the combined Grand Junction/Warwick & Napton/Warwick & Birmingham line, authorised in the four years 1793–96. The Trent–Midlands vertex was to be duplicated by the Leicestershire & Northamptonshire Union Canal, en-Acted in 1793 to run from the navigable Soar (a Trent tributary) at Leicester to Northampton, where it was to connect with a branch from the Grand Junction Canal. But the Union Company ran into difficulties, and their line ended at Market Harborough, having been built from Leicester. The line was completed by the building of the Grand Union Canal, en-Acted in 1810, but conceptually the duplication of this vertex dates from the early 1790's, the period of the 'Canal Mania'.

The fourth arm of the cross, from the Mersey to the Midlands, was duplicated much later, by the Birmingham & Liverpool Junction Canal, authorised in 1826. Its relatively late date is not the only way in which this canal differs from most other English canals, as will be seen.

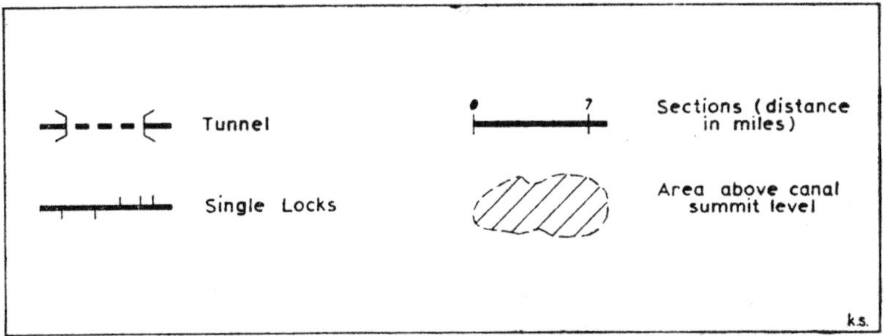

Fig. 3. Symbols Used in Maps of Individual Canals and Summit Levels

The east-west vertices of the figure-of-eight were also multiplied in the early and mid-1790's. The Rochdale and Huddersfield Canals were both authorised in 1794; the Kennet & Avon Canal was authorised in 1794 and the Wiltshire & Berkshire in 1795.

It should be pointed out that this skeletal structure of canals involved only a portion of the total number of canals built.

With the exception of the Birmingham & Liverpool Junction Canal and the Thames & Severn Canal, all the canals which make up the primary and secondary elements of the figure-of-eight network originated respectively in the periods 1766–78 and 1791–95. The second wave of canal promotion owes something to the success of the earlier projects and the time needed to bring them into operation, but the availability of capital and the national economic situation were also important.

WATERSHED CROSSINGS

It is obvious from this outline of the trunk canal network that major watershed crossings are involved on most of the canals concerned. This was in fact a crucial section of the route in two ways.

Firstly, it was important to the success of the canal as a financial venture. Its bearing on the financial aspect of the canal lay in the cost of deep cutting, tunnelling, lockage, and water supply and storage. These features were involved on a summit level to a degree determined largely by the skill of the engineer in selecting the level at which the summit was to be built, and the length for which that level was to be maintained. The smooth movement of traffic at all seasons also depended on the extent to which the natural difficulties of obtaining water on watersheds could be overcome.

Secondly, the choice of location of the summit level was of great importance in influencing a canal's overall route. This becomes apparent if there are considered to be two principal fixed points on a canal—the termini. Once the choice of summit level has been made, two more fixed points, one at each end of the summit level, come into existence, so that the route of the canal and the areas it could serve depended very much on the location of its summit level. (It must also be recognised that the desire to serve certain areas or points would tend to create other more or less fixed points on a general route, thereby partly determining a desirable zone for summit level location.)

For these reasons, attention will now be given to this section of canal routes.

Elevation of the major watershed crossings

Two examples of major watershed crossings will be considered; the Braunston summit level of the Grand Junction Canal and the Oxford Canal summit level (Figs. 4 and 5).

The Braunston summit level crosses the Stratford Avon–Nene watershed and is 3·75 miles long, at an elevation of 357 feet. It is approached from the east by a flight of 7 locks which climb up the west side of the valley of a left-bank headwater of the Nene. The approach from the west is by a flight of 6 locks climbing up the

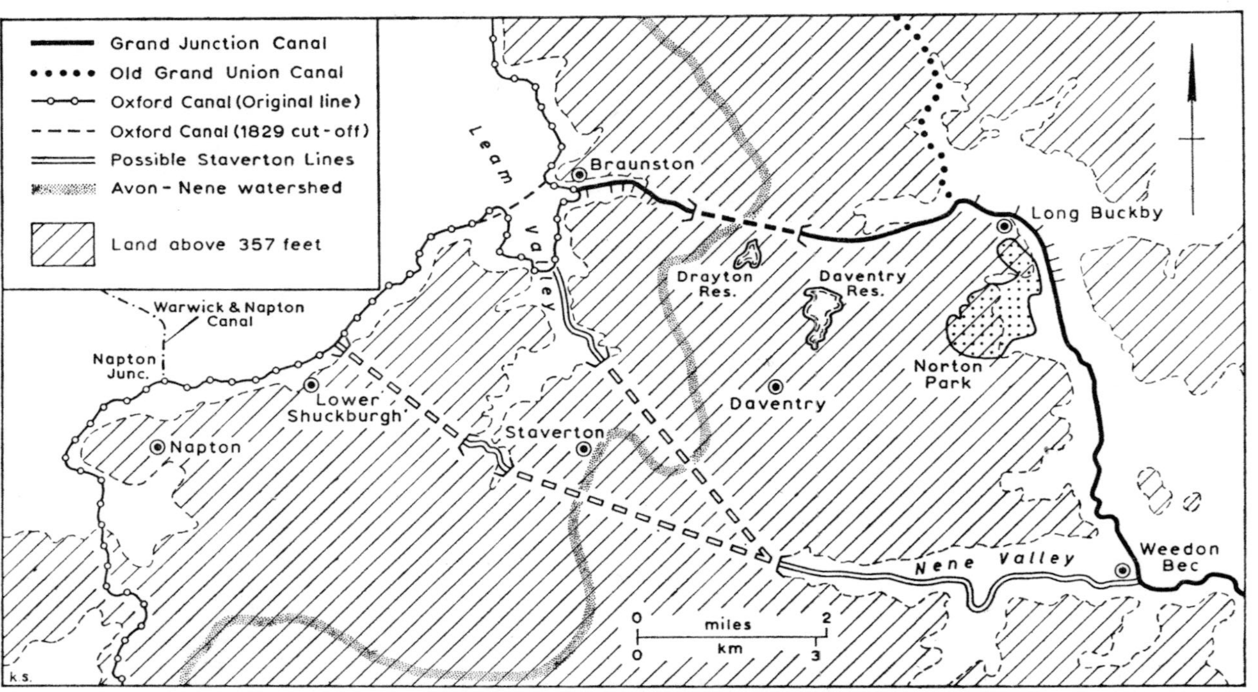

FIG. 4. BRAUNSTON SUMMIT LEVEL, GRAND JUNCTION CANAL

valley of a small right-bank headwater of the Leam. Between them these two headwater valleys approach to within a mile of completely breaching the Jurassic scarp of the Northamptonshire Uplands, here about 500–550 feet high. The intervening col is pierced by Braunston tunnel, 2,042 yards long.

Summit levels were always made as long as possible to provide a reserve of water in the canal; in this case the summit level was restricted at the western end by the position of the Oxford Canal. A junction had to be made with this canal at Braunston, and therefore the fall had to be made to the level of the Oxford Canal east of the junction. At the eastern end of the summit level it seems probable that the limiting factor was the large estate of Norton Park. It may reasonably be assumed, that the level might have been continued further south if Norton Park had not necessitated its descent to a level lower down the valley, which enabled it to skirt the park (Fig. 4).

There appears to be a suitable crossing of the scarp further south, where the head of the main Nene valley penetrates deep into the uplands near Staverton. It suggests a better, shorter route than that actually followed to the north, since the Grand Junction Canal was directed towards a junction with the Oxford Canal as close as possible to Napton Junction and the Warwick & Napton Canal; the ultimate destination was Birmingham. A consideration of what might at first sight appear to be an alternative route demonstrates clearly the sort of considerations that influenced the choice of route and the location of the summit level.

There are at least three factors which weighed against the use of this route.

Firstly, if it is assumed that the summit level at a Staverton crossing would be at a height similar to that of the existing Braunston summit, there would have to be a tunnel at least 1 mile longer than Braunston tunnel. It might be supposed that money saved on the extra 3 or 4 miles of canal north of Weedon Bec—which would be made unnecessary by using the Staverton line—could be used to raise the level of the canal at this point, so reducing the length of the tunnel at the Staverton crossing. In fact some saving in tunnel length could be made by raising the level at Staverton, because of the gentle slope of the land away from the divide, but the distance from the western end of the tunnel to the Oxford Canal would be about 3 miles, so that little saving in distance over the Braunston route would be made as far as the length of canal to be built by the Grand Junction Company was concerned.

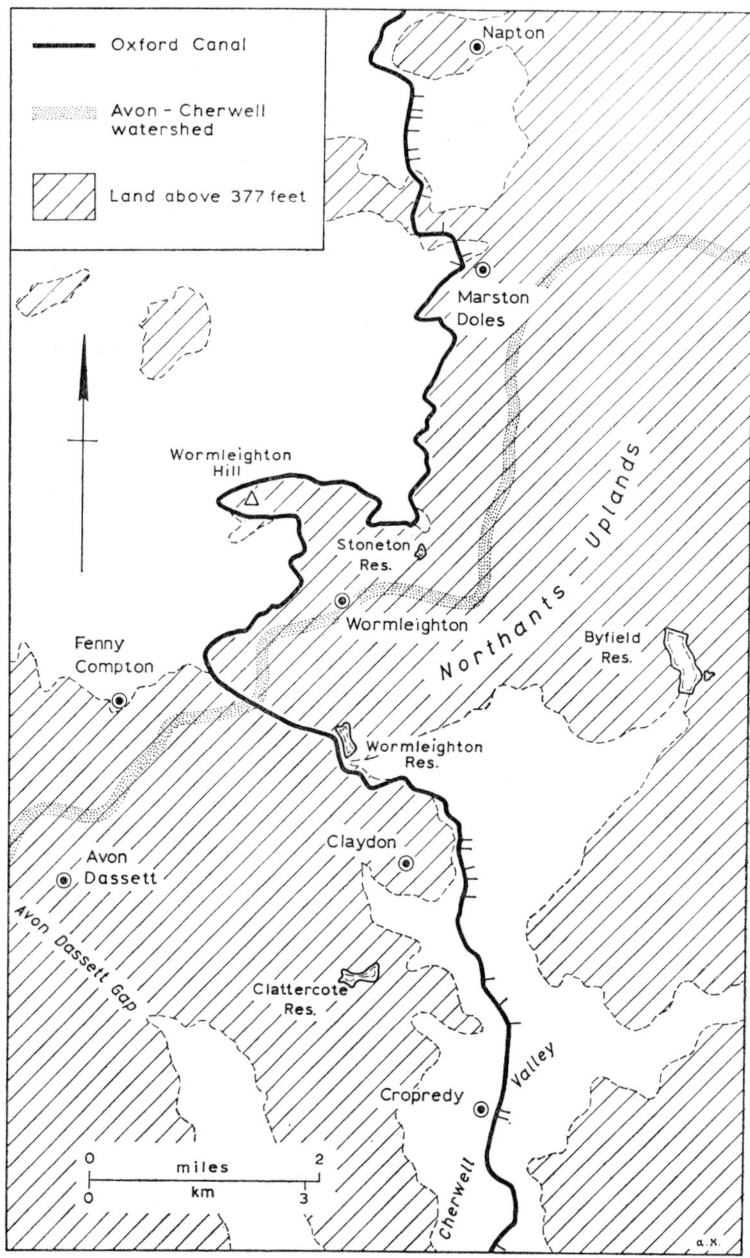

FIG. 5. OXFORD SUMMIT LEVEL

Secondly, the height of the ground above a tunnel near Staverton would be greater than at Braunston, involving greater cost and difficulty of construction.

Thirdly, a direct line to the westward to join the Oxford Canal near Lower Shuckburgh would not be practicable, since it would involve crossing the Leam valley, probably by a contouring detour, and also a deep cutting or even a second tunnel through the western outlying edge of the scarp overlooking the Avon valley. This would be followed by a very steep descent to the Oxford Canal. If this were avoided by directing the line northwards down the Leam valley to join the Oxford Canal near Braunston, the same junction would have been reached as by the northerly line, with the expenditure of considerably more effort and expense.

A simpler example of the preference for lower watershed crossings is afforded by the summit level of the Oxford Canal (Fig. 5). The level is 11·5 miles long, at an elevation of 377 feet. It enables the canal to pass from the Cherwell (Thames) drainage system to the Avon system by using the Fenny Compton Gap to cross the Jurassic scarp trending north-east to south-west. The gap appears to have existed preglacially, and it acted as one of the spillways for water from glacial Lake Harrison impounded in what is now the Avon valley (Bishop, Shotton).

In combination with the Cherwell valley, the gap is a fine routeway through the North Oxfordshire and Northamptonshire Uplands; this routeway is used by the Oxford–Leamington railway line as well as by the Oxford Canal. The floor of the gap is about 405 feet above O.D. Only two other gaps through the Jurassic scarp occur at comparable heights. These are at Braunston and Avon Dassett, at 470 feet and 435 feet respectively. This was the choice that presented itself for an easy route between the Thames and Avon basins. The selection of the Fenny Compton Gap may be regarded as almost automatic, with only the Avon Dassett Gap, less than 3 miles to the south-west, to rival it. The Braunston Gap was well off a direct route from the upper Thames to the Coventry Canal, which was the basic 'desire-line' of the Oxford Canal. It was used later when the Grand Junction Canal was built to cut out the River Thames section of the London–Birmingham navigation.

The Avon Dassett Gap lies at a height about 30 feet above the Fenny Compton Gap, and the length over which this greater height is maintained would have involved a considerable length of tunnel, if the same summit elevation were employed. A higher elevation for the summit would have made it shorter than the one obtained

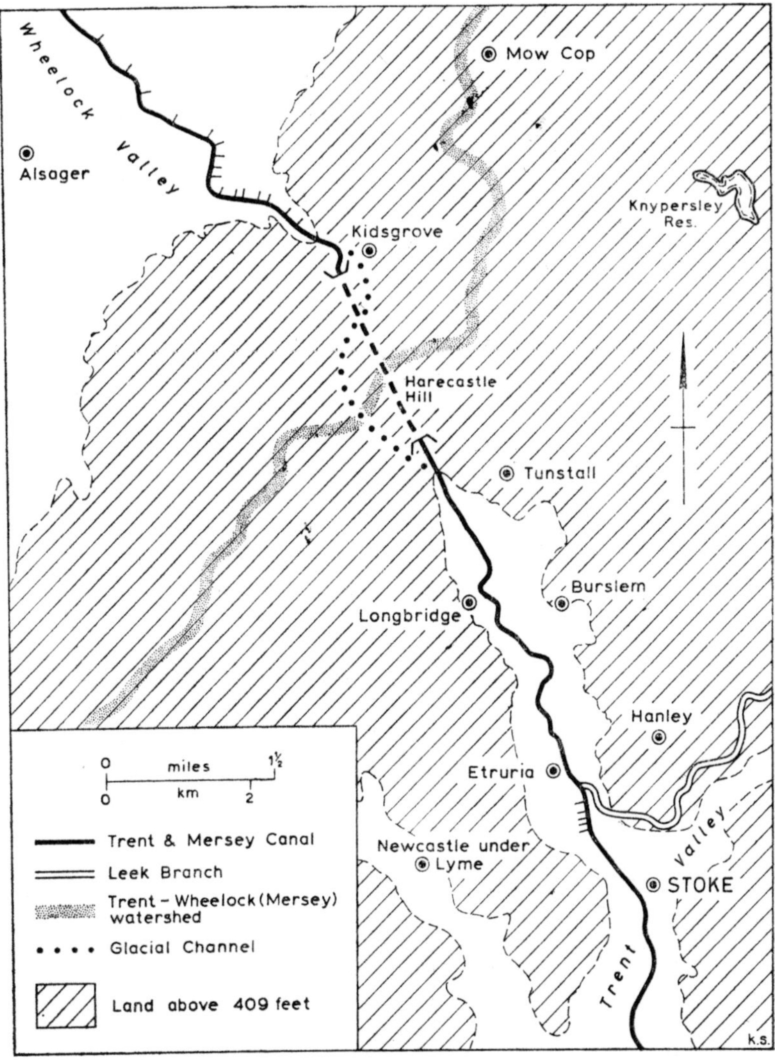

FIG. 6. TRENT AND MERSEY SUMMIT LEVEL

by using the Fenny Compton Gap, and would have formed a smaller reserve of water.

The examples of the Braunston summit level and the Oxford Canal summit level demonstrate the general rule, that the route chosen crosses the main divide between the terminal areas in its lowest part.

There are only two exceptions to this rule out of the sample of eleven canals; the Trent & Mersey and the Leeds & Liverpool Canals. The former crosses the Trent–Mersey watershed in tunnel at a point where it is 105 feet higher than the through glacial overflow channel only a few hundred yards to the south-west (Fig. 6). The reasons for this anomaly were threefold; to enable a long summit level to be constructed, to act as a reserve of water—always a desirable feature of a summit level; to intercept springs by tunnelling and so provide an assured water supply—again a common way of obtaining water; and to connect with existing underground coal workings in the hill, thus supplying further water to the canal, as well as some traffic.

The Leeds & Liverpool Canal avoids a crossing of the Aire–Ribble watershed between Gargrave and Hellifield, at about 525 feet, in favour of a shorter route to the south, which tunnels beneath the divide where it is at an elevation of about 540 feet. In following this route the summit level crosses the watershed three times, which is unique among the canals of the sample (Fig. 7). This is due to the glacial interruption of the local drainage; the summit level follows a trough of low land which is probably of pre-glacial origin, infilled with later alluvium and with an almost level floor. This low-lying area is itself now anomalous in that it crosses the watershed twice. An interesting parallel has been noted in the case of railways across the Great Australian Divide. A similar situation arises with a poorly defined watershed area (Appleton, 1963).

In general it can be said that where more than one route crosses the same major divide, the earlier route or routes follow the lowest crossing of the two or three routes (Table 1).

Apart from the Braunston summit of the Grand Junction Canal, the only exception to the rule is the Birmingham & Liverpool Junction Canal. This avoids the south-westward projection of the Pennines which form the Trent–Mersey watershed, and which the Trent & Mersey Canal crosses at 409 feet; instead it takes an easier line to the south-west across the watershed in the gently undulating Cheshire and West Staffordshire lowlands. Its summit level was designed to join, at the same level, the Staffordshire & Worcestershire summit. The reason for the different directions of the two

TABLE 1—Summit Level Heights

'Primary' Canals (dates of first Acts)	Summit level heights (ft. above O.D.)	'Secondary' Canals (dates of first Acts)	Summit level heights (ft. above O.D.)
Leeds & Liverpool (1770)	487	Rochdale (1794)	600
		Huddersfield (1794)	638
Trent & Mersey (1766)	409	B'ham & L'pool Junction (1826)	341
Staffs. & Worcs. (1766)	341	Worc. & B'ham (1791)	452
Oxford (1769)	377	Grand Junction (1793)	
		Braunston	357
		Tring	391
Thames & Severn (1783)	363	Kennet & Avon (1794)	451

routes, which enabled the later canal to take a lower line, lies in their different functions. These are expressed in their titles; the Trent & Mersey Canal (also called the Grand Trunk) was to connect these two rivers, with the additional aim of serving the Potteries in the Burslem–Stoke area of the Trent valley. The route taken was in these circumstances the most reasonable one. In contrast, the Birmingham & Liverpool Junction Canal was built at the end of the period of canal promotion, in the face of threatened railway competition. It was designed to provide as direct a route as possible between Birmingham and the Mersey, connecting the Birmingham Canal, already being shortened, with the Ellesmere Canal at Nantwich.

Thus, although the two canals cross the same major watershed, their different economic functions meant that their crossing places were well separated; the relatively easy route followed by the later canal would have been unreasonable for a canal designed to link the Trent with the Mersey. In fact, the direct line between the Birmingham Canal and Nantwich lies across a zone where the watershed is not clearly marked by relief obstructions. Its lowness is mainly due to the large-scale glacial diversion of the Severn drainage towards the south. This enabled the Birmingham & Liverpool Junction Canal to achieve its aim of reaching a summit level at the level of the Staffordshire & Worcestershire Canal summit in an unbroken climb from Nantwich, without having an additional summit on the main watershed. A single summit was always the aim of the canal engineer, since each summit level required its own water supply, and increased the quantity of water required (Appleton, 1962, p. 7).

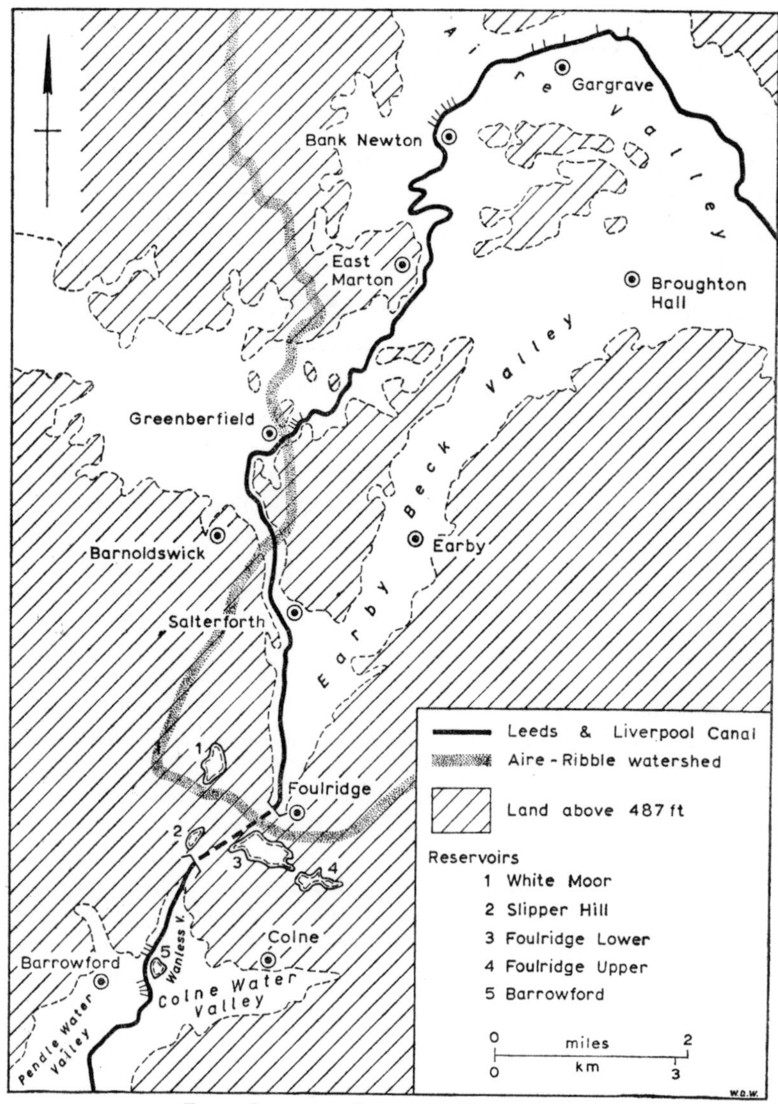

Fig. 7. Leeds and Liverpool Summit Level

The preference of earlier canals for lower watershed crossings is to be expected in view of the relatively low level of engineering knowledge and of capital availability in the early canal era. Later canals were able to overcome greater physical obstacles, in the light of better knowledge and financial backing and also the wave of confidence that prevailed in the Canal Mania years of 1792–94.

Water supply and summit level elevation

Once the location of a summit level was determined by the lowest available crossing of the divide in question, its length and elevation were determined by the interaction of two conflicting factors, and by the compromise made by the engineer. These conflicting factors are the need to secure as large a supply of water as possible and to keep lockage to a minimum, which implies a low summit level, and the need to keep the length and depth of tunnelling and cutting to a minimum, which implies a high summit level. The case for tunnelling or deep cutting was always strengthened by the partial saving in construction costs due to the amount of lockage that would thereby be unnecessary. There would also be a saving in navigating time and in water losses, since locks invariably lose a certain amount of water through leakage. The way in which these conflicts were resolved is exemplified by the Worcester & Birmingham Canal, which had to climb from the Severn at Worcester to the Midland Plateau at Birmingham. The canal's summit level is exceptionally long—14 miles—and is at an elevation of 452 feet (Fig 8).

Like the Birmingham & Liverpool Junction Canal, it was built into an existing canal network, and the northern end of its summit level eventually connected with the main line of the Birmingham Canal at Worcester Bar, near the centre of the city. (The connection was at first prevented (Hadfield, pp. 85–86 and 137)). From this junction southwards the summit level lies on the Triassic rocks of the Midland Plateau south of Birmingham. Where the rim of the Lickey Hills guarding the southern edge of the plateau is encountered south of King's Norton, however, the level is maintained, rather than allowing a higher summit to be imposed by the hills.

This is directly attributable to the reasonable concern that there would be insufficient water available at such a high level. Three tunnels, West Hill (2,726 yards), Shortwood (613 yards) and Tardebigge (580 yards) maintain the level. Nearby, the hills are negotiated by the Lickey Incline on the former Birmingham & Gloucester Railway main line, with a maximum gradient of 1 in 37. The railway passes the hills in cutting, at a maximum elevation of about 600

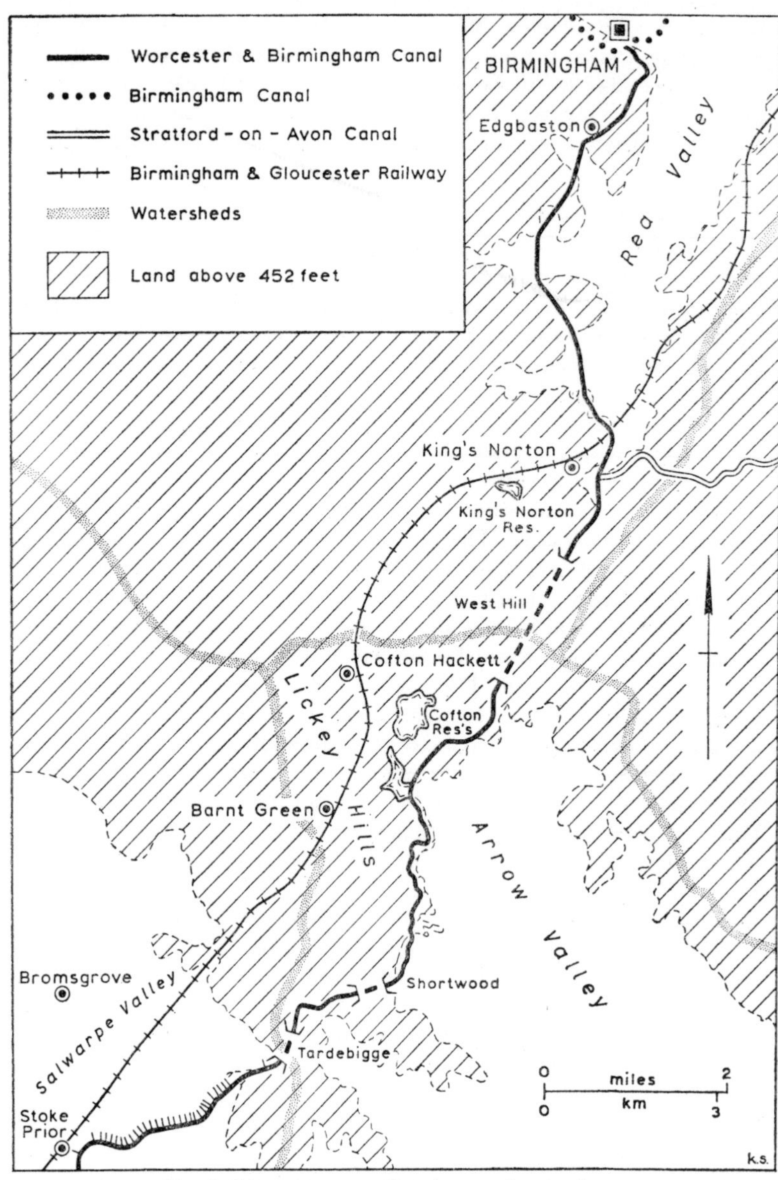

FIG. 8. WORCESTER AND BIRMINGHAM SUMMIT LEVEL

feet near Barnt Green. The canal passes the obstacle at a level more than 100 feet below this, but has to tunnel in the process; the water supply problem prevented the adoption of a summit level with a subsequent descent on to the lower surface to the north, as was carried out on the railway. The southern part of the Lickey Incline corresponds to the steep fall from the summit level at Tardebigge to Stoke Prior, in a flight of 36 locks within 3 miles.

Water for the summit level comes from reservoirs at Cofton and King's Norton. The former store water from headwaters of the River Arrow, a southward-flowing tributary of the Avon, and King's Norton reservoir stores water from a headwater of the River Rea, a northward-flowing tributary of the Tame. The Rea (Tame)–Arrow (Avon) watershed is negotiated by West Hill tunnel, and the Arrow (Avon)–Salwarpe (Severn) watershed by Tardebigge tunnel.

Summit level elevation was thus determined mainly by water supply factors, which operated to make it desirable to lay out a level below the elevation at which adequate reservoirs could be sited and supplied with water without pumping, which added to capital and running costs. If a watershed crossing on the surface would be too high to allow this, the level was carried on through a tunnel or deep cutting where practicable, rather than having recourse to pumping. It should be noted that several canals whose summit levels were designed without pumped water supplies had this facility added later, due partly to leakage and partly to increases in traffic.

On occasion a surface crossing could be made without involving a pumped water supply, but if the fall of the ground away from the divide were so steep that the summit level could be maintained for only a short distance on each side of the divide, then cutting or tunnelling would be carried out to form a longer summit level which could act as a reserve of water.

This practice may be illustrated by the case of the Leeds & Liverpool summit level at Foulridge (Fig. 7). The level is 6 miles long, at an elevation of 487 feet. As has already been noted, it crosses the Aire–Ribble watershed three times, once in Foulridge tunnel (1,640 yards) and twice at ground level near Barnoldswick.

There was some controversy as to the length and elevation of the summit level. Initial proposals were for a level which would cross the Foulridge Gap on the surface (i.e. at about 530 feet), and which would be only about a mile in length. This could have been supplied with water in a similar way to the existing situation, with gravity feed from the two Foulridge reservoirs, and two smaller reservoirs on the northwest flanks of the gap. But it was felt that this would be too short a length to maintain a constant depth of water in the channel while supplying water to the downward lockage on each side. A longer level was therefore sought.

It was found that by tunnelling through the ground at the Foulridge Gap, about 50 feet below ground level at the most (Plate 1), the two levels adjacent to the originally proposed short summit level could be connected. The level would in this way be lengthened to about 6 miles, forming an adequate reserve of water for the lockage on each side. Therefore, although a summit level with a gravity water-feed could have been formed without tunnelling, a tunnel 1,640 yards long was built to extend the summit level and enlarge the water reserves within the canal itself.

The Braunston summit level on the Grand Junction Canal is in the same category (page 4). The Huddersfield Canal summit level displays the same features, since in theory a very short surface summit could have been built and fed with water without pumping while in fact Standedge tunnel was preferred. In practice, however, the height of the Pennine divide would have required such a great lockage, with its attendant cost and navigation time, that the tunnel would probably have been preferred simply to cut out this lockage, even without the need to lengthen the summit level by tunnelling. In fact the level was only 3·75 miles long in any case, most of it being in tunnel (5,415 yards). The canal had a recurrent water shortage; it was under-supplied from the first, the summit level had insufficient reserves for the heavy lockage on each side, and millowners in the Colne and Tame valleys imposed restrictions on the amount of water the Canal Company could use. This conflict in water-use was common during the 18th and early 19th centuries, before steam power was generally applied in previously water-powered mills.

Tunnelling, deep cutting and pumping

In extreme cases the necessity of both tunnelling (or deep cutting) and pumping was either anticipated or very quickly discovered from experience. This occurred on three out of the twelve summit levels examined; on the Kennet & Avon Canal, the Thames & Severn Canal, and on the Tring summit of the Grand Junction Canal. Each will be discussed in some detail.

The Kennet & Avon Canal summit level is 2 miles long, at an elevation of 451 feet (Fig. 9). It crosses the watershed between the Kennet (Thames) and Salisbury Avon systems by means of Savernake

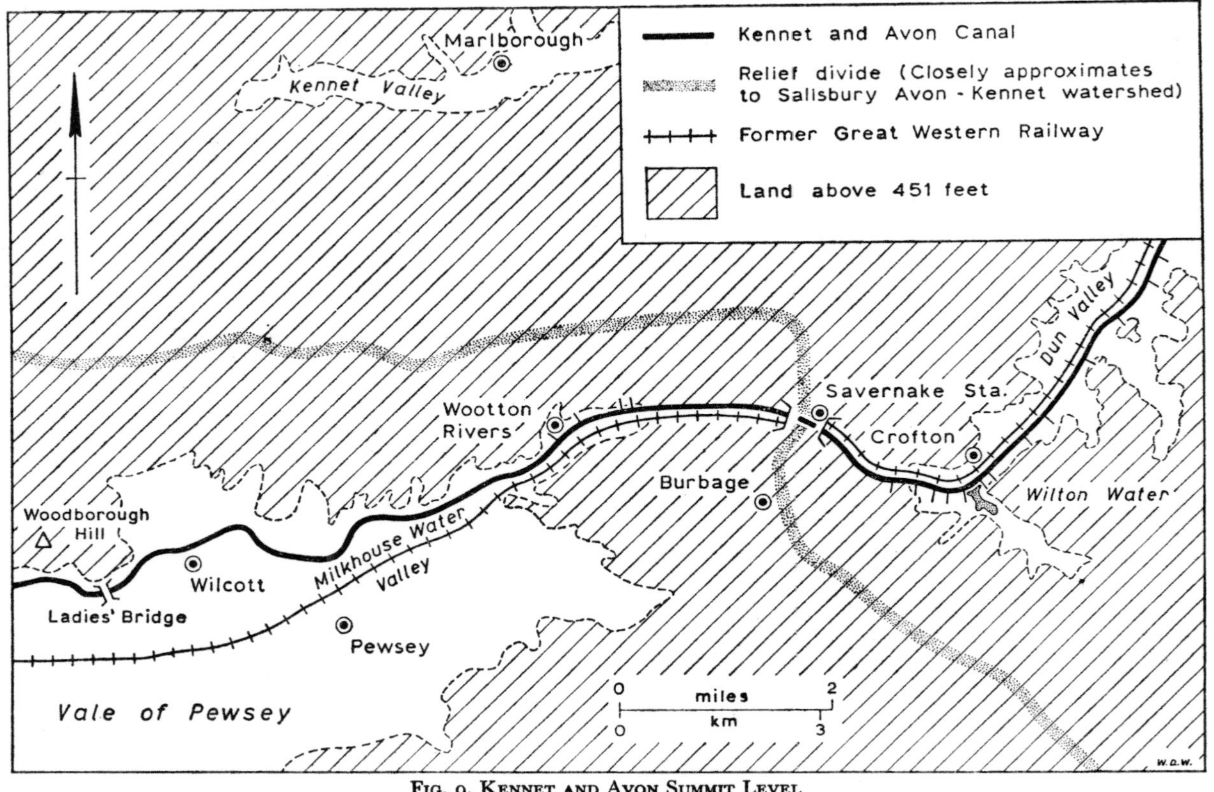

FIG. 9. KENNET AND AVON SUMMIT LEVEL

tunnel which is 502 yards long (Plate 2). The canal makes use of the close headward approach of the valley of the River Dun, which flows into the Kennet at Hungerford, and the valley of the Milkhouse Water, a Salisbury Avon headwater. The summit itself is situated in a relatively low passage in Upper Greensand at the eastern end of the Vale of Pewsey. The through depression affords a route out of the vale used by the main London–Exeter railway line as well as the canal. This geomorphic feature may be a relic of a drainage line flowing along the Vale of Pewsey from west to east, towards the Kennet and the Thames Basin, in which case it would represent a line of drainage similar to the Kennet; Linton suggested the latter is a relic of a west-east drainage pattern once widespread over a large part of South East England (Linton). Problems encountered with the length, elevation and water supply of the summit level were similar to those already explained in relation to other summits, but they were more acute here. The land falls away from the divide area into the dry valley heads on each side, so that a long summit could only have been obtained at the expense of a long tunnel. For example, one proposal during the planning of the canal was for a summit level 5·25 miles long. This would have involved 4,000 yards of approach cutting, and a tunnel 5,368 yards long. The engineering and financial impracticability of such an arrangement left the alternative of a higher-level crossing, with a shorter summit level and tunnel, and less approach cutting. This however raised two problems. The shortness of the level would reduce the water reserve in the canal; and the water supply to the summit would have to be raised from the reservoir at Wilton Water (fed from a spring close to the Upper Greensand—Lower Chalk junction) by pumping, whereas gravity feed would have been possible with a lower level. In the event, the time and expense which would have been involved in the lower level, together with the engineering problems it presented, caused the decision to go in favour of a higher, shorter level with a short tunnel as built. Water was pumped by steam-driven pumps at Crofton (which are still in existence) from Wilton Water and fed into a mile-long feeder which joined the summit immediately west of Crofton Top Lock. To compensate so some extent for the shortness of the level, its channel was built deeper than the rest of the canal—a common practice on summit levels, to increase water reserves.

The Thames & Severn Canal summit level is 8 miles long, at an elevation of 363 ft. It crosses the Thames-Frome (Severn) watershed by means of Sapperton tunnel, 3,817 yds. long (Fig. 10). It was

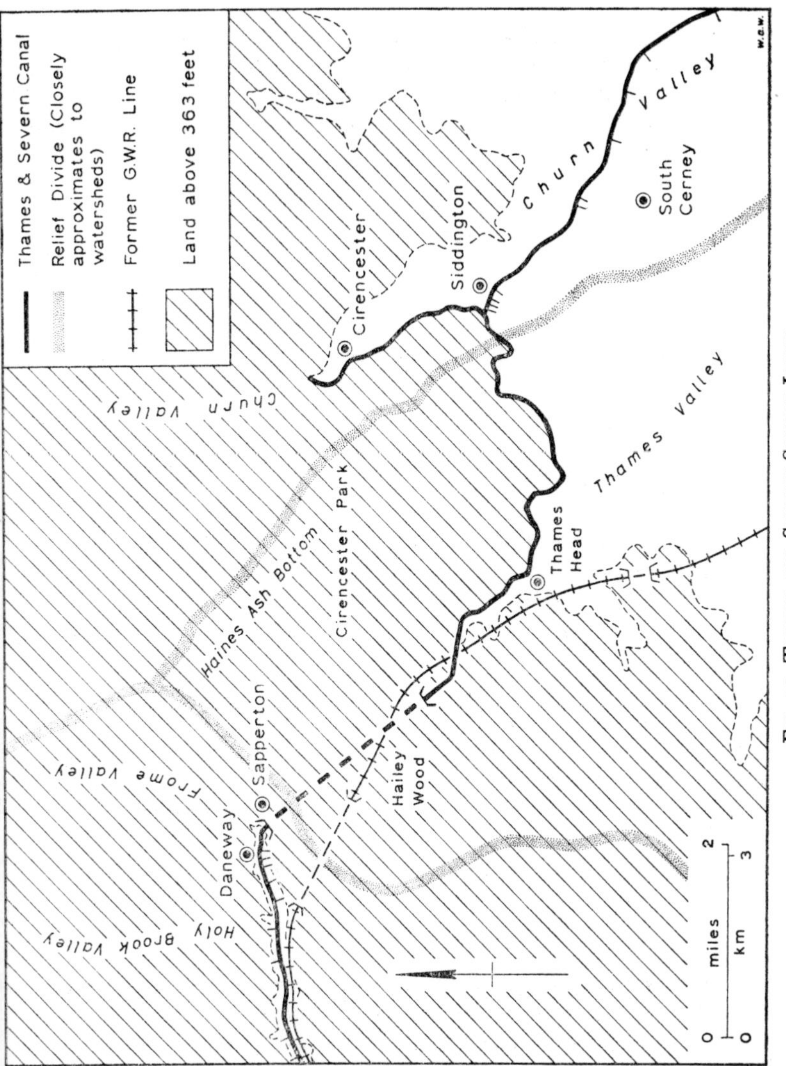

Fig. 10. Thames and Severn Summit Level

Plate I Foulridge Tunnel, southern portal, Leeds & Liverpool Canal
The tunnel portal is visible on the left of the view. The path used by horses to reach the northern end of the tunnel (there being no towing path) passes over the portal. The Colne—Skipton railway line, crossing the Aire—Ribble watershed on the surface, is marked by the line of telegraph poles leading in from the right of the view. Note the shallow depth of the tunnel below ground surface; this length of the tunnel was built on the 'cut and cover' principle.
Looking north-north-west.

deliberately laid out by the engineer, Robert Whitworth, so that it was a few feet below the level of the River Churn at Cirencester. By doing this, he secured a constant supply of water for the summit level without having recourse to pumping. There was such confidence in the Churn as a water source that no reservoirs were built. The level having been determined in this way, it followed that a long tunnel would be needed to negotiate Sapperton hill, and Sapperton tunnel was at that time (i.e. in the 1780s) the longest canal tunnel to be built. That this was accepted as being necessary indicates that Whitworth was well aware that water supply in this area of porous rocks and seasonal springs would be at a premium. Even so, it soon proved impossible to keep pace with the leakage from the summit level in its crossing of the fissured and porous Great Oolite formation. Soon after the canal was opened throughout, in 1789, a wind pump was installed at Thames Head springs to pump water into the summit. This was replaced in 1794 by a Boulton & Watt steam engine.

The Tring summit level of the Grand Junction Canal crosses the Thame–Bulbourne (Thames) watershed in a dry gap breaching the Chiltern escarpment. Its floor is at an elevation of 425 ft. at the highest point. The summit level is 3 miles long, at an elevation of 391 ft. (Fig. 11).

The summit water supply is pumped into four reservoirs situated close to the spring line in the Lower Chalk near the foot of the scarp, and feeds by gravity into the main canal via the navigable Wendover Branch. The summit level was prevented from being in the preferable situation below the spring line by the tunnelling that would have been necessary at this lower level. As it was, over 1 mile of deep cutting (35 ft. deep maximum) was necessary to maintain the level at 391 ft., mostly cut through unconsolidated Valley Gravel which floors the mouth of the gap. Unless this cutting had been made, the level would have been higher and shorter, presenting problems of water supply on two counts. Firstly, a summit length shorter than the existing one would have provided an insufficient reserve of water in the canal itself, and secondly, water would have had to be pumped to a greater height than under the existing arrangement. Thus the elevation of the summit level at Tring was influenced by a desire to maintain it as low as possible, to lengthen the level and provide an adequate reserve of water, and to reduce the cost of pumping, without at the same time having recourse to tunnelling.

It is significant that all these cases where pumping was necessary,

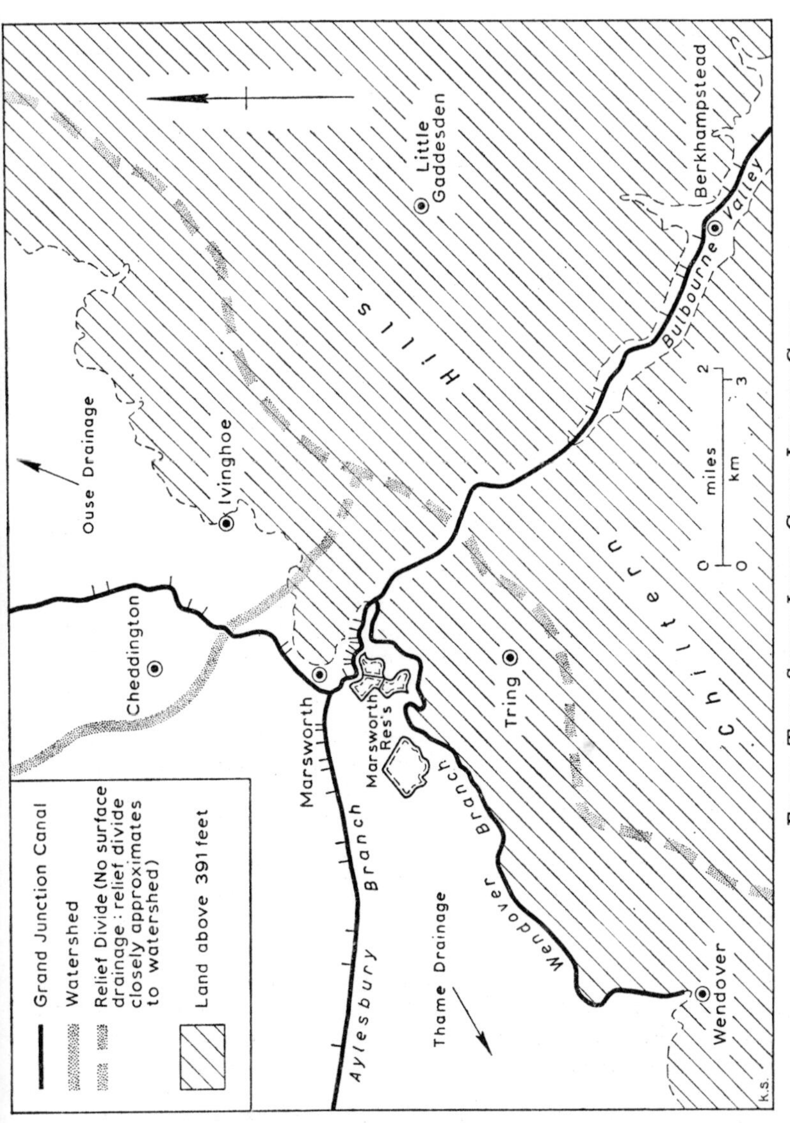

FIG. 11. TRING SUMMIT LEVEL, GRAND JUNCTION CANAL

PLATE 2 SAVERNAKE TUNNEL, EASTERN PORTAL, KENNET & AVON CANAL
The tunnel, also known as The Bruce Tunnel, is 502 yards long. The water level is about 1·5 feet below working level. Note the railway signal box above and to the right of the portal.
Looking north-east.

either from the beginning or very soon after opening, are situated in Chalk or Oolite uplands, characterised by the scarcity and irregularity of surface water flow, and by high permeability. Of the canals considered here, no other summit level is situated on such rocks. The scarcity of surface water in these areas was obvious enough, and was allowed for in the arrangements made for water supply. But the high permeability of these rocks was not well understood at this time, and as a result the two Thames–Severn canals in particular suffered from water losses.

Geological factors

As a summary of the lack of geological knowledge of the 18th century, the remarks of William Smith, a pioneer of stratigraphic geology, are of interest. They occur in a report he made to the Kennet & Avon Canal Company in 1812 on the subject of water losses and embankment slips on the canal (BTHR[1], KAC 4/12).

> '. . . everyone must have allowed . . . [the Kennet & Avon Canal] . . . to have been executed in that masterly manner in which Mr. Rennie's works are always executed. But the making of canals previous to the execution of the Kennet & Avon and Coal Canals[2] was a new thing in all the southern and western part of the Island, and never before attempted in any of these strata where the greatest difficulty has been experienced except in the tunnel and summit level of the Thames & Severn Canal. The summit level of the Kennet & Avon Canal is very fortunately made through much better strata than that of the Thames & Severn . . . the great difficulty of making watertight Sapperton tunnel and a summit level of 9 miles . . . [was] . . . as I conceive one of the reasons why that canal has so long remained unprofitable . . .'

Geological factors did not generally affect canal routes, since the state of knowledge of geology during the period of canal construction was not advanced. A similar lack of accurate information existed as regards hydrology: few records of stream-flow or water table levels existed, and such phenomena as landslips and seasonal springs were not better understood until the early 19th century. As a result, potential water supply was often over-estimated, and geologically unsuitable terrain was crossed. Routes were only affected to a small extent by the 'geological factor' in cases where the surface materials were known to be unsuitable for canal construction. For example, the Kennet & Avon Canal east of Hungerford keeps to the alluvium-covered floor of the Kennet valley,

[1]BTHR British Transport Historical Records, Paddington.

[2]The Somersetshire Coal Canal, engineered by Smith himself.

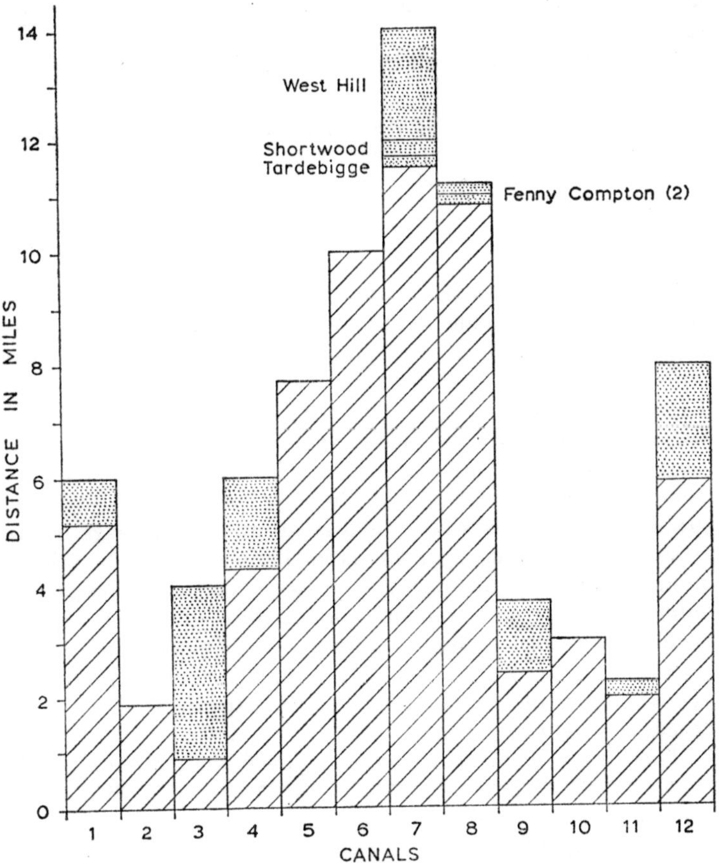

14
12
10
8
6
4
2
0
DISTANCE IN MILES
West Hill
Shortwood
Tardebigge
Fenny Compton (2)
1
2
3
4
5
6
7
8
9
10
11
12
CANALS

INDEX TO CANALS

1 Leeds & Liverpool
2 Rochdale
3 Huddersfield
4 Trent & Mersey
5 Birmingham & Liverpool Junc.
6 Staffs. & Worcs.
7 Worcs. & Birmingham
8 Oxford
9 Gr. Junc. (Braunston)
10 Gr. Junc. (Tring)
11 Kennet & Avon
12 Thames & Severn

Length of summit tunnels
Total summit level length

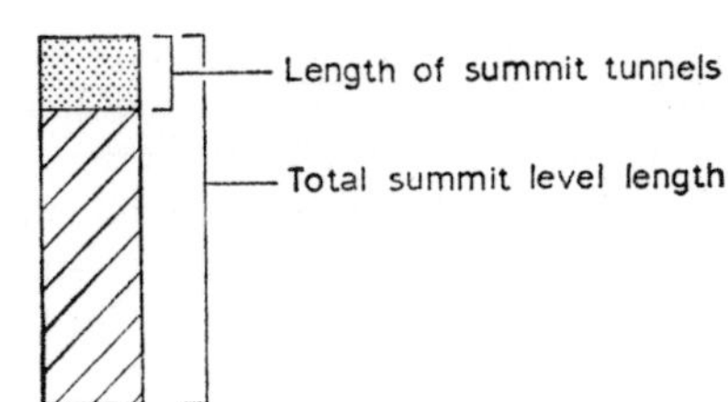

FIG. 12. LENGTHS OF SUMMIT LEVELS AND SUMMIT TUNNELS

deliberately avoiding the porous surface materials along the valley sides.

Whitworth's remark (BTHR, KAC 3/12) on the subject in 1789 was:

> 'From my knowledge of the country I know there is a necessity to keep near the river course, and in the lowest ground, upon account of the nature of the soil [elsewhere], which is very bad for holding water.'

Summit level lengths

Finally, as a summary of the factors discussed in this Chapter, reference may be made to the diagrammatic representation of the lengths of summit levels and tunnels (Fig. 12). Of the four canals without tunnels on their summit levels, two (the Birmingham & Liverpool Junction and the Staffordshire & Worcestershire) made use of the same gently graded crossing of the Trent–Severn watershed in the Tettenhall Gap. This gap is a relatively easy routeway along the western edge of the Midland Plateau; it existed in pre-glacial times, and was later used as an overflow passage by water from ice-dammed lakes in the Trent Basin (Wills). The other two summit levels without tunnels (the Rochdale Canal and the Tring summit of the Grand Junction) both use through-valleys which owe their present form mainly to the action of glacial meltwater. In addition, both are relatively short summit levels, suggesting that the avoidance of tunnelling was at the expense of shortening the levels, as would be expected.

There is a wide diversity in the lengths of the summit levels, from under 2 miles (Rochdale) to 14 miles (Worcester & Birmingham). This should be regarded as a result of physical restrictions and the influence of landforms acting through the media noted—tunnelling, lockage, cutting and water supply—rather than the result of deliberate intentions; the aim was always to obtain as long a level as possible, for the sake of water reserves.

The exceptional length of the Worcester & Birmingham summit level is attributable on the one hand to the extension of a surface with relatively subdued relief south of Birmingham, and on the other hand to the need to avoid creating a higher summit level in the Lickey Hills, which brought the response of heavy engineering works and three tunnels on the summit level.

CHAPTER II

Analysis of Morphology based on Morphological indices

IN order to proceed further in generalising about relationships between canal morphology and the physical environment, more precise measurement of canal morphology is required. These numerical measures should reflect as closely as possible actual variations in morphology, and in order to reduce the weighting effect of a few extreme values on a set of data, the number of values should be large for each length of canal. Also required is as wide a coverage of the canals as possible, using uniform methods of measurement.

Two morphological indices are used:

(i) an index of sinuosity;

(ii) a mean figure derived from the height of each embankment and the depth of each cutting.

Using these indices, two general hypotheses may be tested:

(*a*) canal morphology as measured in this way varies with landform;

(*b*) canal morphology varies with the advance in engineering knowledge during the period of canal construction.

The hypotheses will be considered further below (see p. 61).

MEASUREMENT OF THE INDICES

Sinuosity Index

The sinuosity index measures the degree to which canals deviate from a straight line on a small scale; that is, involving deviations of the order of a mile or less in each direction. It is important to recognise the scale of the index and its implications. It is difficult, if not unjustifiable, to make generalisations on the basis of a measurement of large-scale deviations from an assumed or inferred desire line. Canal promotion usually involved the issue of prospectuses listing towns through which the line was to pass, but no indication is given in contemporary sources regarding the accuracy of these lists as bases for constructing desire lines. The Leeds & Liverpool Canal is a significant exception, in that post-Act route changes produced abundant material indicating the reasons for these changes, and this has been discussed elsewhere (Farrington, 1970), but even here the original route, much of which was unaffected

---

Plate 3 Oxford Canal Summit Level between Stoneton and Marston Doles
The canal traverses a gently-sloping surface stretching from the foot of the Jurassic scarps, a lower portion of which forms the horizon. This unshortened section of the canal follows the contour of this sloping surface very closely. Note the normal situation of the towing path, on the 'outside' of the channel.
Looking east.

by these changes, is like other canal routes in being open only to conjectural analysis on the basis of desire lines. The problem that is difficult to resolve is to what extent the prospectus lists were compiled *after* the general route had been determined, possibly by factors of water supply or constructional difficulty. In other words, is there any justification in assuming that the desire-lines ran between the places listed? Without more evidence there seems little justification for accepting this as a basis for the measurement of large-scale deviations.

There are also difficulties in measuring small-scale sinuosity, but they can be overcome satisfactorily. Suppose it were decided to adhere as closely as possible to the actual canal course, but to retain the element of a straight 'desire-line' base. It might be proposed that sinuosity could be measured as shown in Diagram A, Fig. 13. The straight lines *ab*, *bc*, *cd*, are each of length one unit—say 1 mile—and the sinuosity is measured by the lengths *ab*, *bc*, *cd*, measured along the line of the canal.

Although this reflects moderately well the small-scale sinuosity, it is based on the assumption that the ideal route would have been as defined by the linked straight lines *ab*, *bc*, *cd*. This assumption is not well founded, and cannot be adequately supported in practice.

An alternative method is to adopt the line of the canal itself as the base from which its sinuosity is to be measured. This is the more logical approach, since it is desired that measurements should be derived as far as possible directly from the line of the canal as built, and not from an assumed or theoretical 'desire-line'. The procedure is thus resolved into marking off points unit distance apart along the line of the canal, and measuring the straight line distance between consecutive points, as in Diagram B, Fig. 13. Lengths *ab*, *bc*, *cd*, measured along the canal are of unit length, and sinuosity is measured by the straight line distances *ab*, *bc*, *cd*. It is apparent that measured sinuosity here depends on the positions of points *a*, *b*, *c* and *d*. In an extreme case the result would be a misrepresentation of the true situation, as in Diagram C, Fig. 13. The canal line is *abcd*, with straight reaches connected by sharp bends. This canal's sinuosity expressed by the straight line distances *ab*, *bc*, *cd*, would appear as zero—the canal would be represented as having no deviations from a straight line. Although this is an extreme case, some misrepresentation from this source is likely.

To reduce this misrepresentation a modification is made. As before, unit lengths are measured along the line of the canal, but intermediate points are added mid-way along each unit length

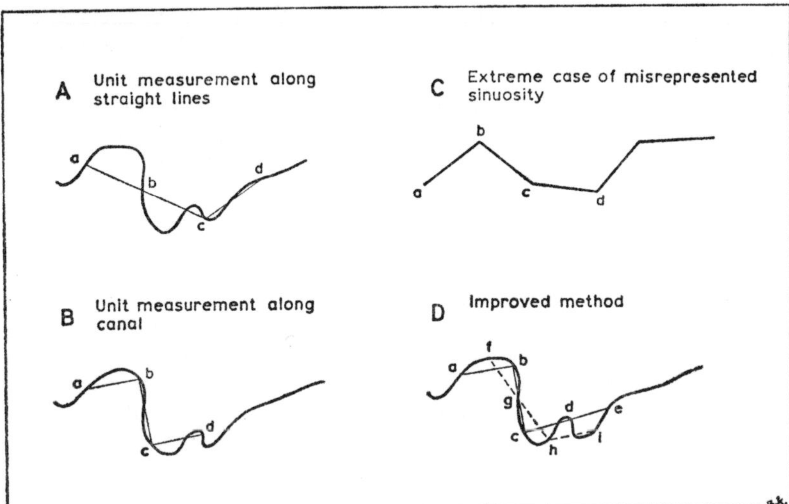

FIG. 13. MEASUREMENT OF SINUOSITY

measured along the canal, as in Diagram D, Fig. 13. Points, *a*, *b*, *c*, *d* and *e* are unit length apart measured along the canal. The intermediate points *f*, *g*, *h*, and *i*, are also unit length apart, and canal length $af = fb = bg = gc = ch = hd = di = ie$. The sinuosity value of canal length *bc* is then given by:

$$\frac{1}{2}\left(\textit{straight line bc} + \frac{(\textit{straight line fg})}{2} + \frac{(\textit{straight line gh})}{2}\right)$$

Similarly the sinuosity value of canal length *cd* is given by:

$$\frac{1}{2}\left(\textit{straight line cd} + \frac{(\textit{straight line gh})}{2} + \frac{(\textit{straight line hi})}{2}\right)$$

In this way succeeding sinuosity values are 'smoothed' and extreme variations are avoided. Of course any number of intermediate points could be used. This adds proportionately to the work-load in measurement. The larger the cartographic scale at which sinuosity values are measured, the more sensitive is the index in reflecting each deviation in the course of a canal. Therefore the largest practicable scale should be chosen; for this work the 2½ inches to 1 mile O.S. maps were used. Using this method a series of values is obtained for each canal, each value representing the sinuosity of one mile of canal. The smallest series is for the Huddersfield Canal,

21 miles long, and the largest series is for the Leeds & Liverpool Canal, 127·5 miles long. 2·5 inch units were used, so that a value of 2·50 indicates absolute straightness; the lower the value below 2·5, the greater the degree of sinuosity. This body of data can be used in comparing one canal with another or one length of canal with another.

Cutting and Embankment Indices

The ideal measure of these features would be the volume of earth moved, since this would most directly have reflected the work and cost involved. Such information was not available without large scale field surveys. Next in order of suitability is the depth and height of each cutting and embankment. A source of such information exists in data gathered by the British Waterways Board, in 1964–65. The survey was carried out on all the Board's waterways in accordance with a uniform and detailed programme. Cutting less than 15 ft. deep was not recorded, whereas all embankments were recorded. This is not a problem in the present context since cutting data are not compared directly with embankment data. Comparisons are restricted within the two sets of data, and as the 15 ft. lower limit applies to all cutting data, consistency is maintained.

The figure used in the morphological analysis is maximum depth/height of each cutting/embankment. The data are reduced into the framework adopted in the measurement of sinuosity. The same unit lengths along the canal are used, and the cuttings and embankments whose mid-point falls within each length are tabulated. The figure for use in analysis is calculated as the mean of the maximum depths/heights of the appropriate cuttings/embankments located within each mile length on the canal. A series of values is obtained, one for each mile along the canal, except for the miles where no mid-points of cuttings or embankments occur within the mile.

It would be desirable to evolve a method of measuring landforms according to slope angle and slope sinuosity (i.e. 'contour sinuosity'), for correlation with the indices of canal morphology. What is required is a set of measurements for the tracts of country through which the canals pass, on the lines of morphometric analysis. This suggests a line for further research; the methods would be applicable to road and rail lines as well as canals.

In the absence of a numerical description of landform, the procedure adopted has been the less objective one of subdividing the canals into lengths according to their passage through landforms

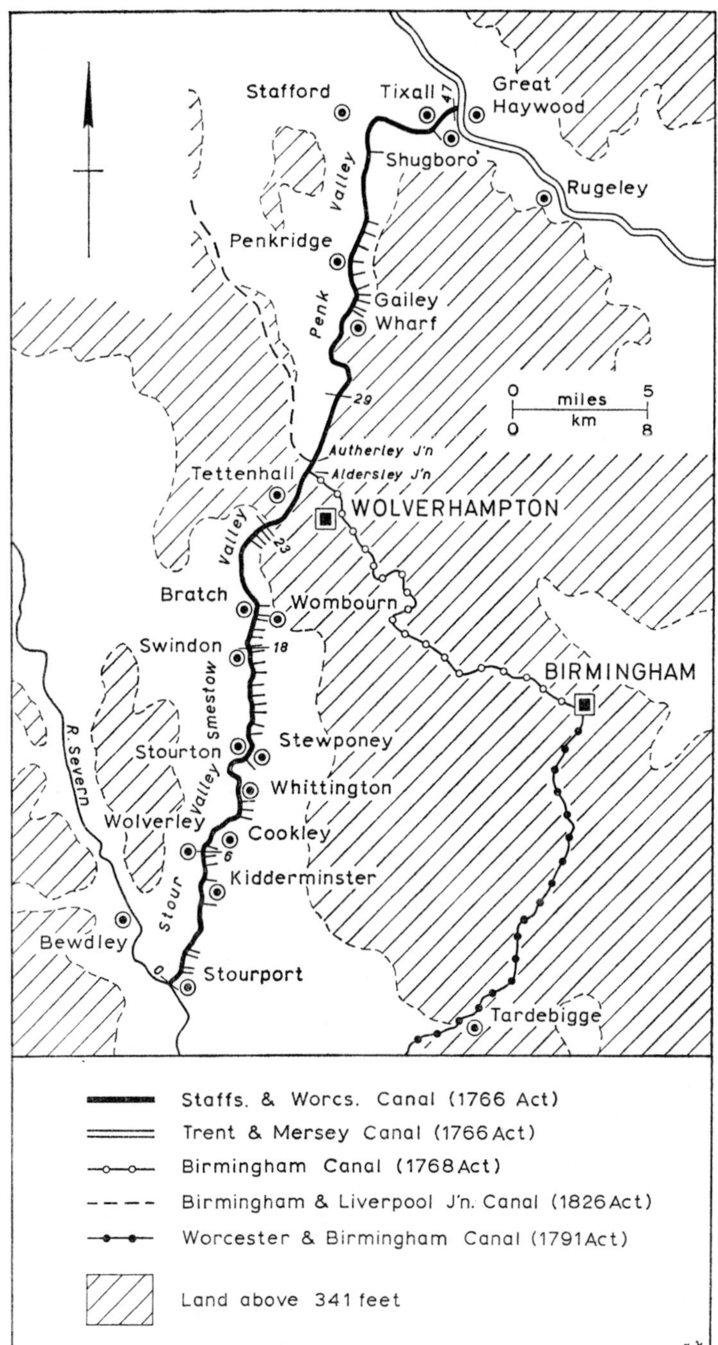

FIG. 14. STAFFORDSHIRE AND WORCESTERSHIRE CANAL-ROUTE

which are judged to be likely to differ from length to length with respect to the requirements of canal construction. The extent to which canal morphology, as expressed in indices of sinuosity and cutting and embankment, is affected by these landform variations may then be assessed within this framework, and generalisations derived. The small figures at intervals along the Staffordshire & Worcestershire Canal in Fig. 14 are examples of this procedure. The figures represent mileages, and correspond to the divisions made in the way described. The resulting canal 'sections' are used in statistical analysis (see Chapter 3). The principal analytical procedure, however, is based on the 'morphological profiles' as exemplified in Fig. 16.

Example of analysis using morphological indices

Before making generalisations about canal-landform relationships the morphology of the Staffordshire & Worcestershire Canal will be described and analysed using the morphological indices. This serves as an example of the analysis on which the generalisations are based.

The Route. The Staffordshire & Worcestershire Canal crosses the Trent–Severn watershed by following the Penk and Stour–Smestow Brook valleys, which are linked by the Tettenhall Gap west of the Midland Plateau (Fig. 14). Between Stourport, where it joins the Severn, and Kidderminster, the canal follows a terrace of the Stour, but north of Kidderminster the valley becomes narrow and steep-sided. The canal is closely restricted to the valley, and its morphological features derive largely from this influence. North of Stewponey the north-south line of the Stour valley is continued by the Smestow Brook. By following this valley and its headwaters, the canal reaches the level of the trough overlooked on the west by a scarp of Lower Keuper Sandstone, on which stands Tettenhall itself. To the east is a more gentle rise towards Wolverhampton, on the Coal Measures Midland Plateau.

The long summit level (10 miles), at a height of 341 ft., is situated in the poorly drained area between the Smestow Brook and Penk valleys, and is continued into the Penk valley itself. This was carried out for the sake of water supply, which comes from reservoirs at Gailey (Fig. 15).

From Gailey there is a gradual fall along the open Penk valley to the Sow valley at Baswich, which is followed to the junction with the Trent & Mersey Canal at Great Haywood.

Landforms and canal morphology: Long profile. Although the general form of the long profile is that of two river valleys with a long summit

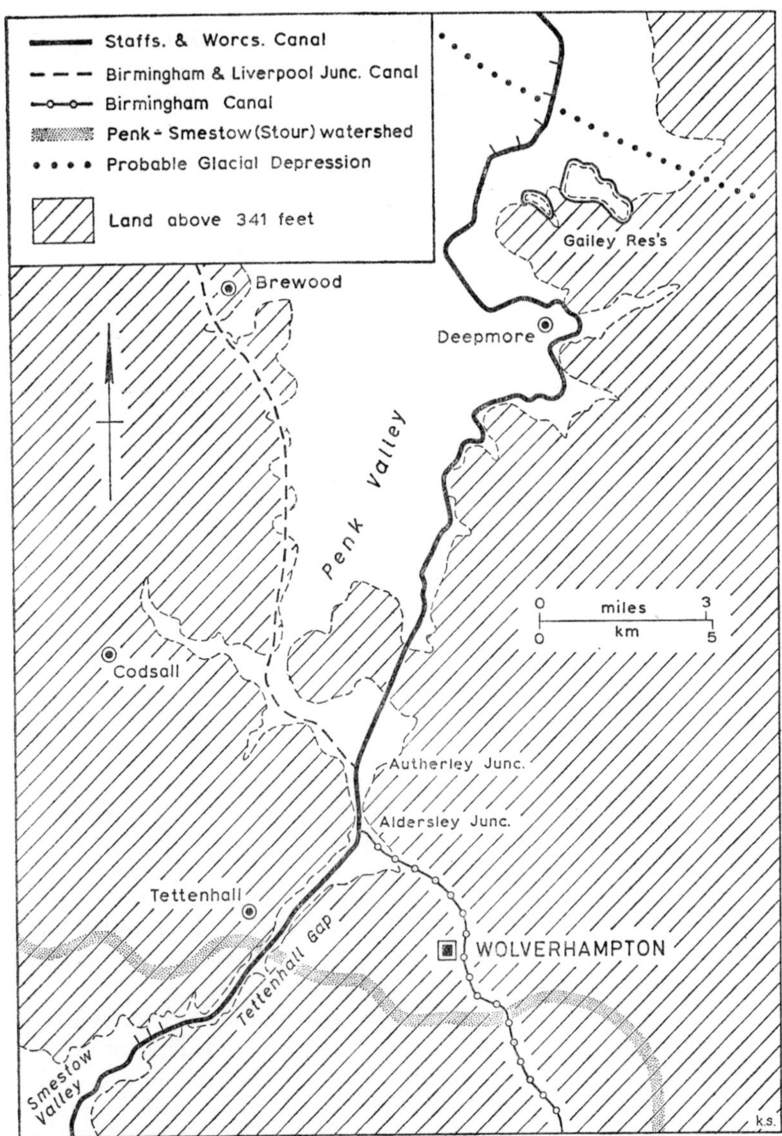

FIG. 15. STAFFORDSHIRE AND WORCESTERSHIRE CANAL—SUMMIT LEVEL

level connection, several elements may be recognised within this framework (Fig. 16).

From Stourport to Kidderminster (miles 0–4) there is a gentle rise of only 3 locks in 4 miles. This section follows the wide terraced Stour valley. North of Kidderminster the rise is steeper but gradual, up the Stour and Smestow valleys as far as Swindon. North of this point the Smestow Brook describes a wide arc towards the west, which the canal avoids by climbing more steeply across the intervening spur. This is reflected in the increased gradient seen on the profile between miles 17 and 21. There are 10 locks on these 4 miles compared with 13 locks on the 13 miles between Kidderminster and Swindon.

At the highest point of the intervening spur, the canal is above the level of the Smestow Brook to the west, and there follows a level pound almost 2 miles long between this point and the rejoining of the Smestow Brook valley south of Tettenhall. Here occur the last 3 locks to the summit level, which passes through the Tettenhall Gap and along the upper slopes of the Penk valley. At Gailey a minor watershed between the Penk and a small right-bank tributary is crossed, and the canal falls along the tributary valley through 10 locks in 5 miles to a level a few feet above the Penk floodplain at mile 39. From this point to Great Haywood Junction the fall is very gentle through only 2 locks in 8·5 miles. This reflects the low gradients of the Penk and Sow valleys.

Landforms and canal morphology: Sinuosity. The sinuosity profile may be conveniently analysed in the same sections as those adopted for the long profile (Fig. 16).

For the first 4 miles north of Stourport the level valley floor allows moderately high sinuosity values to be maintained. North of Kidderminster the Stour valley is quite deeply cut and sinuous. Moreover, it is seldom more than a few hundred yards wide, so that this section of the canal, as well as the section in the similar lower Smestow Brook valley, is characterised by a high degree of sinuosity (low values, about 1·90). Straighter lengths correspond to less sinuous lengths of the valley.

At Swindon (mile 17) where the canal climbs out of the sinuous Smestow valley onto a less-dissected surface, a series of higher values begins, indicating a lower degree of sinuosity. This continues through the Tettenhall Gap, itself markedly linear. The Penk (Trent)–Smestow Brook (Severn) watershed is crossed in the Gap at mile 26.5, but instead of falling to the north from the watershed area the canal maintains its summit height for a further 7 miles in order to

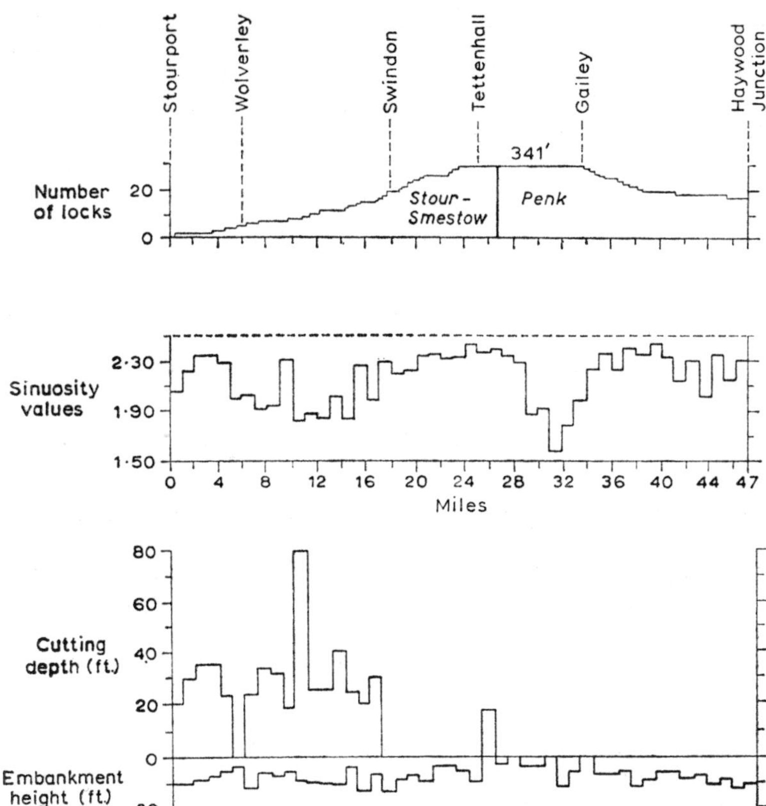

FIG. 16. STAFFORDSHIRE AND WORCESTERSHIRE CANAL—MORPHOLOGICAL PROFILES

secure a water supply at Gailey. The maintenance of the level involved a sinuous contouring detour, which could have been avoided if a lower level had been adopted. The sinuosity profile reflects this highly sinuous length (miles 29–34).

On the section which falls relatively steeply down the minor right-bank tributary valley of the Penk (miles 34–39) a uniformly low sinuosity is maintained; the valley is open and imposes few restrictions on the line. North of Lower Drayton, where the main Penk valley is joined, sinuosity is variable. A relatively high degree of sinuosity (low value) occurs on mile 41–42 (2·13), where the canal avoids the Penk floodplain; on mile 43–44 (2·01) where the line swings sharply from a northward heading to the south-south-east at the confluence of the Sow and the Penk; and on mile 45–46 (2·13) where a Z-shaped bend is necessary to cross the Sow by Aqueduct. This crossing is required for two reasons. Firstly, to join the Trent & Mersey Canal the Trent must be crossed; to cross it below its confluence with the Sow, where it is appreciably wider, would be more difficult. Secondly, Shugborough Park extends along the right bank of the Sow, and along the Trent below the Sow–Trent confluence. It is probable that difficulties would have arisen if the Company had attempted to pass through the Park.

The physiographic differences between the deeply incised Stour-Smestow valleys and the open vale of the Penk are visible in the sinuosity profile of the canal. The high degree of sinuosity associated with the northern end of the summit level at Deepmore is not relevant in this connection; it is related to water supply, as has been pointed out. Apart from this length, the degree of sinuosity of the Stour-Smestow section of the canal (miles 0–24) is generally higher (lower values) than that of the Penk–Sow section (miles 34–47). The mean value of sinuosity on the former (including the high values of the upper Smestow Brook–Tettenhall length) is 2·11, compared with 2·30 on the latter. Excluding the Smestow Brook–Tettenhall length, the mean value of sinuosity on length 0–17 is only 2·05.

The inferred link between canal morphology and landforms is that the open vale topography north of the watershed enables a straighter course to be maintained. On the southern length of the canal, rejuvenated valleys tributary to the Severn impose stricter control over the line. These physiographic contrasts have been attributed to the rejuvenation begun by the glacial diversion of the upper Severn into the middle Severn, which has been

> . . . primarily responsible for the deep etching of the trunk and tributary valleys of the middle Severn which is in such strong contrast to the open vales of the Trent side of the watershed of England. (Wills).

A similar relationship is seen in the Cutting and Embankment profile.

Landforms and canal morphology; cutting and embankment. The most striking feature of the cutting profile is the deep cutting that occurs between Stourport and Swindon, and the complete absence of cutting over 15 ft. deep on the rest of the canal, apart from the watershed crossing. The deep cutting in the Stour–Smestow valley is necessitated by the narrowness of the valley; in order to avoid the river and its mills, the canal had to be cut into the steep valley sides. In 8 of the 17 miles the mean maximum cutting depth is 30 ft. or more.

As significant as the presence of deep cutting on the Stour–Smestow valley section is its virtual absence north of Swindon (mile 17). The valleys followed north of this point are open, with easily graded sides. Nowhere is it necessary to cut into the valley sides to accommodate the canal. The watershed crossing too is remarkably easy in this respect, due to the comparatively gentle fall of land on each side of the divide; a long summit level could be preserved without recourse to cutting or tunnelling.

The embankment height profile is less strongly influenced by landforms. It is generally uniformly low, with few values greater than 10 ft., and none greater than 14 ft. A slight embankment height reduction is noticeable on the summit level, where the canal follows the bottom of the dry Gap. It is to be expected that embankment would be a minor feature close to the watershed crossing, where the crucial factor is normally the depth of the canal level below the land, rather than its height above it.

The low embankments of the Penk–Sow valleys are associated with the gently-sloping valley sides which the canal traverses. But the similarly low embankments of the Stour–Smestow valleys call for comment in view of the exceptionally deep cuttings on this section (these are often rock cuttings in Bunter Pebble Beds Sandstone). It is obvious that a channel was made by cutting into the valley sides rather than embanking out from the sides. The reason for this was probably the width of the valley, insufficient to allow the canal to be built out from the side. This would therefore limit embankment height in the Stour–Smestow valleys.

From this analysis of the morphological features of the canal it can be seen that they reflect closely the landforms in the area traversed by the canal. There is a particularly clear contrast between on the one hand the deeply cut and sinuous Stour-Smestow valley section, and on the other the less sinuous, lightly embanked Penk–Sow valley section. This is directly related to the physiographic contrast between the rejuvenated Severn tributaries and the maturely graded Trent tributaries.

This type of analysis carried out on the other ten canals also indicates clearly that contrasts between landform types traversed by the line of a canal are almost always reflected in the indices of morphology. More detailed conclusions also emerge from the analysis.

'Landform difficulty' and canal morphology

A clear example of the relationships between 'landform difficulty' and canal morphology at different periods is afforded by the contrast between the original line of the north Oxford Canal and the shortened line of the 1820's (Plates 3 and 4).

The Oxford Canal was built under an Act of 1769 and opened throughout in 1790. The northern portion of its line, between Fenny Compton and the original junction with the Coventry Canal at Longford, was particularly sinuous, and was laid out in the early 1770's. In the 1820's the threat of a grand canal link between London and the Midlands, by-passing the Oxford Canal, induced the Oxford Company to shorten its line north of Wolfhampcote, under an Act of 1829. The contrast between the two lines summarises the progress made in engineering techniques between 1770 and the 1820's, as well as indicating the new desire for increased speed and efficiency of movement of goods.

The changes made in the shortening may be illustrated in several ways. The sinuosity profile, in Figure 17, is one of these. The

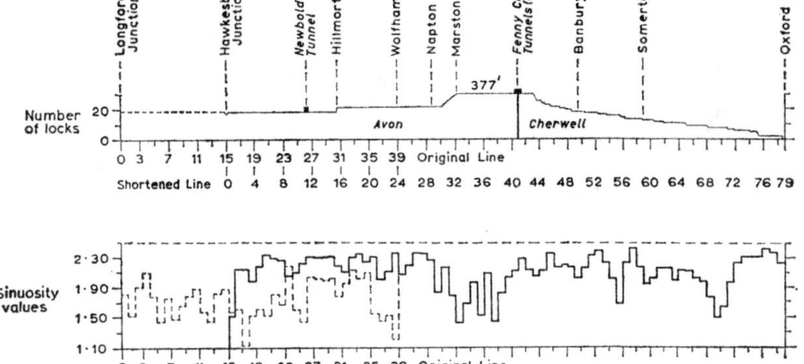

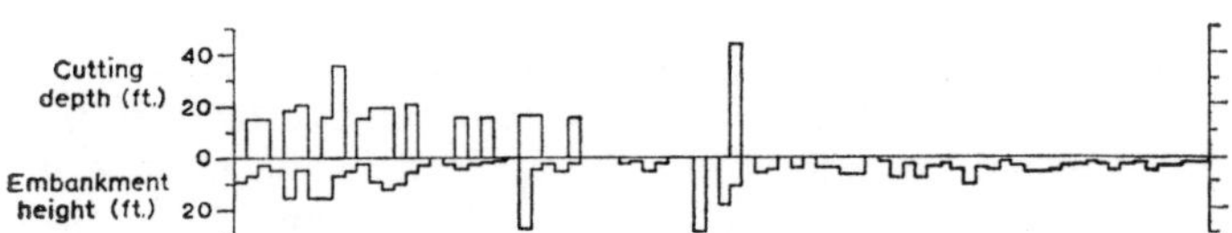

FIG. 17. OXFORD CANAL—MORPHOLOGICAL PROFILES

sinuosity profile of the shortened line is much less irregular and has higher values than the original line, indicating a lower degree of sinuosity. Whereas 19 out of 39 of the 'original' miles had sinuosity values below 1·70, and only 9 of the 39 had values above 2·00, only one mile out of the 24 on the shortened line is below 2.00. This is at Hawkesbury, where no shortening work was carried out under the 1820's scheme because agreement could not be reached on a new junction with the Coventry Canal. The junction at Hawkesbury did not replace the Longford Junction until the late 1830's.

Another method of demonstrating the effects of the shortening is shown in Figs. 18 and 19. Consider first the whole length of the

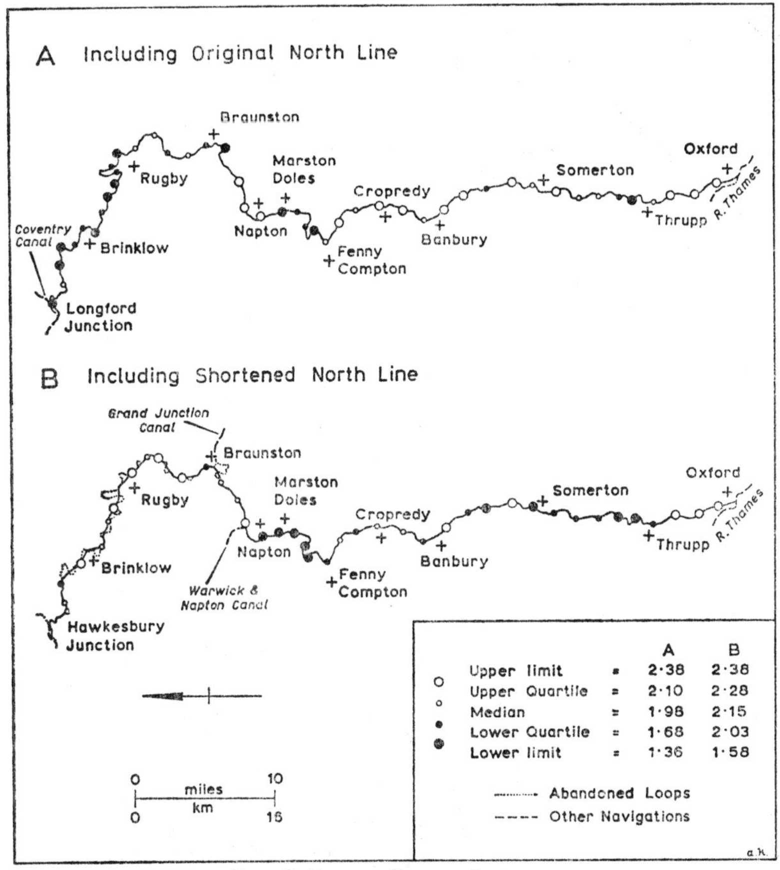

FIG. 18. OXFORD CANAL—SINUOSITY

PLATE 4 OXFORD CANAL AT WOLFHAMPCOTE
A section of the canal built as part of the shortening work of the 1820's. The directness of this line contrasts with the earlier sinuous line. Looking east.

canal (Fig. 18). It is apparent that before the northern section was shortened it possessed almost all of the unit lengths of below-median sinuosity values. Only the summit level along the dissected Jurassic scarp, and two lengths in the sinuous Cherwell valley (at King's Sutton and Shipton) had similarly low values. But after shortening had been carried out on the northern section, it had only two such lengths; while the whole length along the scarp between Napton and Fenny Compton, and most of the Cherwell valley length below Banbury, were made up of units with below-median sinuosity values.

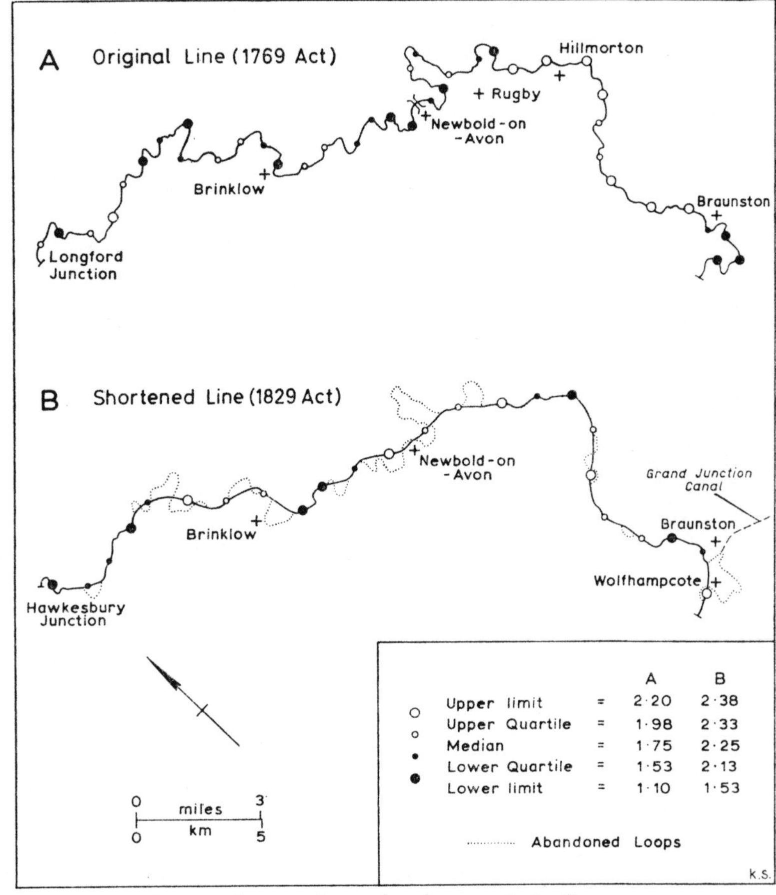

FIG. 19. OXFORD CANAL (northern section)—SINUOSITY

Next consider the section north of Wolfhampcote as an entity (Fig. 19). It is clear that on the original line the re-entrant detours give rise to almost all of the below-median sinuosity values. After shortening, the below-median values are transferred to the lengths of the original line which were incorporated unaltered with the new line. The shortening is also reflected in the higher median and quartile values (i.e. lower degree of sinuosity) on the shortened line.

In order to examine cutting and embankment on the two lines, it is first necessary to justify an assumption made to overcome the absence of cutting and embankment data on the now disused sections of the original northern line. It would be expected that the original line would differ little with respect to cutting and embankment from the rest of the canal within the Avon basin, for which information *is* available—that is, between Wolfhampcote and Fenny Compton. The justification for this statement is based on the argument that the high sinuosity of the original northern line was the means of avoiding significant earthworks, just as large earthworks on the potentially difficult section between Wolfhampcote and Fenny Compton were also avoided by high sinuosity. Therefore it is reasonable to see the cutting and embankment values of the line between Wolfhampcote and Fenny Compton as being comparable with those on the original northern line. It is therefore possible to compare the cutting and embankment values of the shortened line with those of the line between Wolfhampcote and Fenny Compton, and to see this as a comparison between the different engineering approaches of the 1820's and the 1770's respectively.

Between Wolfhampcote and Fenny Compton only three cuttings are recorded, and of these the deepest (43 ft.) is at Fenny Compton, on the site of the tunnels in the Gap, which were opened out between 1865 and 1870. Embankments on this length are generally no more than 6 ft. high. The exceptions are an embankment 30 ft. high at the extreme tip of the outlying Wormleighton Hill (mile 37–38), and embankment 12 ft.–19 ft. high along the steeper scarp slope near Fenny Compton (miles 39–41).

The cutting and embankment profiles of the shortened line are in strong contrast to the Wolfhampcote–Fenny Compton section. Thirteen of the 24 miles on the shortened line have cutting, and on mile 8–9 this reaches a depth of 35 ft. Embankment is similarly generally higher than on the Wolfhampcote–Fenny Compton length, with a maximum of 28 ft. on mile 23–24. Almost all these larger earthworks are associated with the newer cuts. It is significant that most of them are 'true' cuttings or embankments. That is, *both* sides of the canal are in cutting or on embankment, as opposed to the situation on the Wolfhampcote–Fenny Compton length, which is the typical 'contouring' arrangement with towing path embankment and offside cutting where necessary. This situation, which is common on all the canals, is shown in Fig. 20. The engineer's aim was to have the amount of earth excavated to form the channel equal as nearly as possible the amount required to form an embankment for the towing path. The towing path was always on the side shown in the diagram, except where landowners specified that the path should be on the other side, away from their estates, to prevent intrusions by bargemen (Plate 5). In order to accommodate the towing path in the normal position, the canal engineer selected a line traversing a moderate slope, especially on earlier canals.

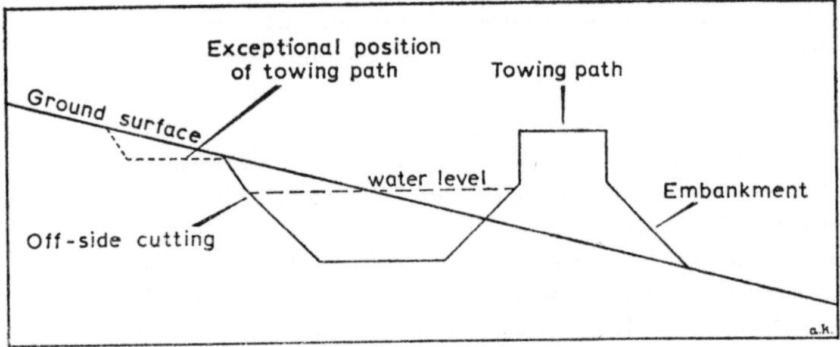

Fig. 20. Ideal Canal Cross-Section

Finally, it should be pointed out that there was an exception to the general rule of low embankment on the original line, in the form of Brinklow Arches, where an embankment and aqueduct (max. ht. 30 ft.) carried the *original* line across the valley of the Smite Brook. This was probably necessitated by the refusal of the owner of the large house and estates of Newbold Revel to allow a 'contouring' detour up and down the Smite valley to pass through his grounds. Such a detour, to negotiate the valley without an aqueduct, would have passed close to the house on three sides. Although a contemporary reference to this has not been traced it would explain the anomaly of the 12-arch aqueduct.

The example of the north Oxford Canal illustrates a tendency apparent in the other canals of the sample studied; in earlier canals an increase in 'landform difficulty' would be overcome firstly by an increase in degree of sinuosity and only in severe cases by a large additional increase in cutting and embankment.

In contrast the tendency on later canals was for physical obstacles to be overcome by an increase in cutting and embankment rather than in degree of sinuosity.

An extreme case of 'landform difficulty' is seen on the section of the Trent & Mersey Canal along the north side of the Weaver Valley west of Northwich. The River Weaver meanders in a wide floodplain bounded on the north by bluffs rising steeply about 100 ft. from the valley floor. The situation is particularly difficult west of Anderton, and it was necessary to avoid the steep bluffs by tunnelling through two spurs of high ground, in Saltersford and Barnton tunnels. This involved individual cuttings up to 70 ft. deep, and individual embankments up to 60 ft. high. These exceptional earthworks were required not only on the tunnel approaches but also on sections where the valley side is extremely steep, involving deep cutting on the offside and high embankment on the towing path side.

The problems of construction were so great that at one stage it was doubted whether they could be overcome. Josiah Wedgewood, the owner of the Etruria pottery factory, and a principal promoter of the canal, wrote in 1774,

> Mr. Henshall deeming it impracticable to make the canal along the high sloping banks on the side of the Weaver, has cut out these new tunnels [Saltersford and Barnton] and very happy it is for us that the ground has been found capable of admitting this alternative, or I verily believe we must have given up this part of our canal, and even now we have some tremendous gullies and sidelong banks to pass over, but none I hope impracticable. (Hadfield, 1966, p. 32)

Thus, extremes of 'landform difficulty' were overcome both by a high degree of sinuosity and by heavy cutting and embankment, since the one without the other did not suffice to overcome the extreme of physical obstacles.

Sinuosity and gradient

The Caen Hill flight of 17 locks at Devizes, on the Kennet & Avon Canal, is a clear example of a low degree of sinuosity (high values) associated with a steep and steady climb (Plate 6). The flight makes the steep ascent from the Bristol Avon valley in the west to the Vale of Pewsey to the east. Where such a gradient had to be overcome, the usual practice of spacing locks at least 100 yds. apart had to be discarded. This spacing enabled water moving into the pound from the lock above to be absorbed in the pound without overflowing, and allowed a lockful of water to be taken from the pound into the lock below without lowering the water level too far for navigation. In the case of the Caen Hill flight, the pounds between the locks were effectively 'lengthened' by providing side-ponds; the volume of the pound was therefore maintained. Other examples of a less extreme type are the section of the Rochdale Canal between Manchester and Failsworth, and the Tardebigge flight of locks on the Worcester & Birmingham Canal.

The corollary of this situation is the usually high and variable degree of sinuosity (low, variable values) which results from the maintenance of a level for some distance. The Thames & Severn Canal summit level exemplifies this statement (Fig. 10). The reasons for maintaining this level for 8 miles have been discussed (p. 18). It results in sinuosity values ranging from a maximum of 2·20 to a minimum of 1·83 (excluding the straight tunnel). The contrast between these values and the sinuosity of the rest of the canal's length, which is involved with steady climbs from west and east towards the summit level, is considerable. The minimum value on the rest of the canal is 2·18, and the maximum is 2·45. The mean value of summit level sinuosity, excluding the tunnel, is 2·03; the mean value of sinuosity on the rest of the canal is 2·35. This represents a strong contrast in terms of the sinuosity index.

The maintenance of a level may involve considerable cutting and embankment as well as a higher degree of sinuosity, as on the Worcester & Birmingham Canal summit level (Fig. 8 and p. 13); or the high degree of sinuosity alone may be the way in which the level was maintained as on the Oxford Canal summit level (Fig. 5, p. 8, and Plate 3). The latter was maintained for 11.5 miles by using a highly sinuous line, when a straighter line at a lower level could have been followed; this, however, would have resulted in a short summit level at Fenny Compton, with an inadequate reserve of water in the canal.

Thus, the maintenance of a level for some distance usually gives rise to a high and variable degree of sinuosity; while a steep and steady climb through flights of locks is accompanied by a relatively low degree of sinuosity, since small relief variations which are potential causes of a high degree of sinuosity are accommodated within the climb.

Cutting, embankment, and major watersheds

The directness of the relationship between cutting and major watersheds is clearly seen on the Grand Junction Canal (Fig. 21). There are five main sections on which cutting occurs: miles 1–5,

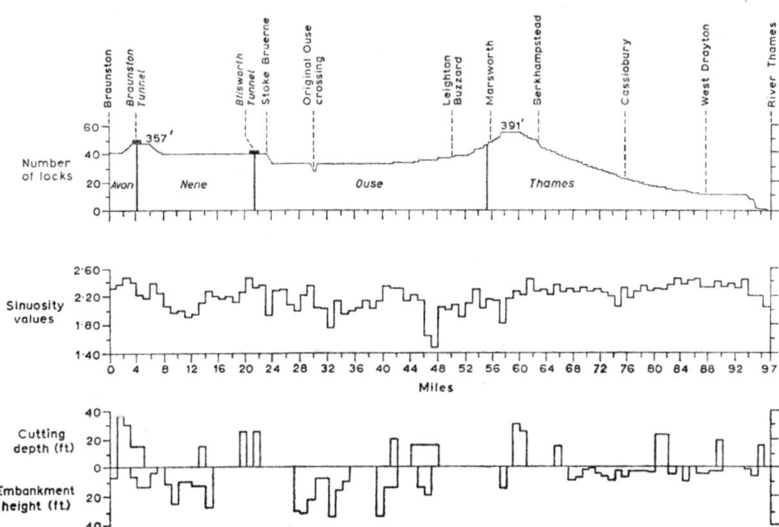

FIG. 21. GRAND JUNCTION CANAL—MORPHOLOGICAL PROFILES

19–22, 41–48, 59–61, and 80–82. Three of these five sections are situated on major watershed crossings—respectively the Avon–Nene (1–5), the Nene–Ouse (19–22), and the Thame–Thames (59–61). Of the remaining two sections, one (41–48) corresponds to the passage of the canal along the Ouzel valley through the Lower Greensand hills north of Linslade, and may be attributed to the steep sides of the valley in this section, necessitating cutting to enable the channel to be formed. The other main length of cutting (80–82) occurs in the Colne valley, in a section where the canal is 'pinned' against the east side of the valley by the Colne; the cutting avoids the need for crossing the river and its floodplain.

This pattern of deep cutting on the major watershed, with otherwise few significant occurrences, is complemented by the embankment profile, which shows a marked tendency for high embankment to occur towards the centres of the two river basins crossed by the canal, the Nene and the Ouse. This accounts for two of the four main 'peaks' on the profile (miles 8–15 and 27–35). A third peak at Fenny Stratford (miles 39–42) is caused by embankment required to maintain a direct course across a wide embayment in the side of the Ouzel valley. The fourth peak is again on the section which passes through the Lower Greensand north of Linslade, and is accounted for by the steepness of the valley sides.

Thus, cutting indicates clearly the relationships between canals and major watersheds; embankment is also a useful index in the analysis of canal—landform relationships.

Canal morphology and the function of valleys as routeways

The contrasts between the lengths of the Oxford Canal in the Avon valleys and in the Cherwell valley illustrate the importance of the 'route function' performed by a valley or drainage basin in influencing canal morphology. In the Avon valley, the canal *traverses* the headwaters of the drainage system, while the function of the Cherwell valley is to provide a *linear* route aligned along the main valley itself.

This difference in the functional relationships between canal and landform is reflected in the morphological profiles (Fig. 17). The long profile in the Avon basin is characterised by the long levels with only two breaks—at Hillmorton locks and between Napton and Marston Doles, where the climb to the summit is made. The long profile of the canal in the Cherwell valley has strong similarities to that of a graded river, with gradient decreasing down-valley. (The relatively sharp descent at mile 75 marks the point where the canal falls to the Thames flood-plain west of Oxford, having left the Cherwell valley proper above the town).

The sinuosity profile indicates a less clear contrast between the lengths of canal in the two drainage system. Even so, it is apparent that the profile of the original northern line is more variable and has more low values than that of the Cherwell valley length. The former has 10 values of 1·50 or less, the latter has only 4.

The cutting and embankment profiles in the two drainage basins differ strongly. The occurrence of relatively deep cutting and high embankment in the Avon basin, and of no deep cutting and no high embankment (max. ht. 12 ft.) in the Cherwell valley, also accords with the situation to be expected from the functions performed by the two drainage systems. The canal in the Avon basin is 'contouring' round the head of the basin, while in the Cherwell system it is following the main drainage line. (It should be noted that the contrast is accentuated by the shortening works on the canal in part of the Avon basin section).

Thus the 'route function' of a valley is important in terms of canal morphology and the overall route.

Engineering capability and canal-valley relationships

The Oxford Canal in the Cherwell valley and the Staffordshire & Worcestershire Canal in the Stour valley (p. 30) are two of the many examples of a situation where the long profile and the sinuosity of the canal are closely governed by the long profile and sinuosity of the valley followed by the canal. Deep cutting and high embankment is infrequent in such cases, unless the valley sides are so steep and the floor so restricted that the channel must be cut into the valley side.

Therefore although there are many lengths of canal where the variable of engineering capability strongly affects the canal—landform relationship, there are situations where this variable is relatively unimportant. These situations arise where canals are situated in valleys the morphological characteristics of which closely control the line of the canal and its gradient; only details are variable.

Optimum situations for canal construction

On the Trent & Mersey Canal below Alrewas in the Trent valley the degree of sinuosity is low (high values of 2·25 or more) with little variation, and embankments are low (10 ft. or less). The canal follows a line which keeps it well away from the river and above its flood level.

Similarly on the Kennet & Avon Canal below Hungerford in the Kennet valley sinuosity is uniformly low, and there is no cutting over 15 ft. deep, and no embankment over 10 ft. high. The canal occupies a situation in the river valley similar to that of the Trent & Mersey below Alrewas.

Therefore, according to the indices of sinuosity and cutting and embankment, the optimum landform situation for canal construction is approached in these lengths of the Trent and Kennet valleys, where there are large continuous level stretches of terraces and floodplain.

Canal-river 'level crossings'

Within the sample of canals considered, the cases of canal-river 'level crossings' and of the intermittent use of a river as the navigable channel are on the Oxford Canal in the Cherwell valley (Frontispiece), on the Kennet & Avon Canal in the Kennet valley, on the Grand Junction Canal in the Colne and Gade valleys, and on the Trent & Mersey Canal at Alrewas.

All but the last of these cases occur in the valleys of dip-slope streams on chalk or porous limestone. (On the Trent & Mersey Canal at Alrewas, a crossing of the wide marshlands of the Trent by aqueduct and embankment would have been costly and difficult.)

The significance of this 'localisation' of level crossings lies in the evenness of the régimes of such streams. They pass through areas of porous rocks which regulate run-off and tend to prevent the extremes of flood and drought of other rivers which made level crossings features to be avoided where possible. The disadvantages of the possible interruption of canal navigation by such occurrences are obvious. The relatively low silt-content of streams in areas of porous rocks would also reduce the danger of obstructions to navigation due to silting.

PLATE 5 LADIES' BRIDGE, KENNET & AVON CANAL
The bridge, ornamented to accommodate the landowner's wishes, is 3 miles west of Pewsey. A tablet in the structure is inscribed 'Erected 1808'. The towing path was placed on the bank from which the view is taken, the bank on the 'inside' of the slope-traversing canal, again at the landowner's request. The natural ground-slope is from right to left across the canal. Note that the material excavated from the channel still formed an embankment on the 'outside' bank, where the towing path would have been more easily and normally located.
Looking west.

CHAPTER III

Further Morphological Analysis: Comparisons between Canals

IN the last chapter examples were used to illustrate conclusions made using morphological data presented chiefly in the form of 'profiles'. To take further the analysis of canal morphology it is necessary to use other forms of presentation and analysis.

GRAPHICAL COMPARISONS

The first stage in this further analysis involves a comparison between graphical presentations of the distributions of sinuosity values, cutting depths and embankment heights for each canal. The distributions of values of the three morphological measures on the sample canals is indicated by the use of quartiles, in Figures 22, 23 and 24.

Basic similarities

Two basic observations may be made at the outset. Firstly, sinuosity values tend towards a value of 2·50, or straightness. This is to be expected, since it accords with the fundamental concept that straight lines of communication are the ideal situation, from which deviations are made only for specific reasons. The Birmingham & Liverpool Junction Canal and the Oxford Canal (including original northern line) are exceptions which will be considered below. Secondly, the distributions of cutting and embankment values tend towards zero. This again accords with the ideal situation with no cutting or embankment. From these observations it may be concluded that a strong fundamental similarity between the canals exists on the basis of sinuosity values (with the two noted exceptions) and on the basis of cutting and embankment. This at least implies that similar factors operated throughout the span of time and space covered by the construction of the canals. It is suggested that this is due to the basically uniform principles of construction which governed the morphology of all the canals (the Birmingham & Liverpool Junction excepted).

The two canals already noted as being exceptional in terms of sinuosity values are the Birmingham & Liverpool Junction Canal and the Oxford Canal including the original northern line. The

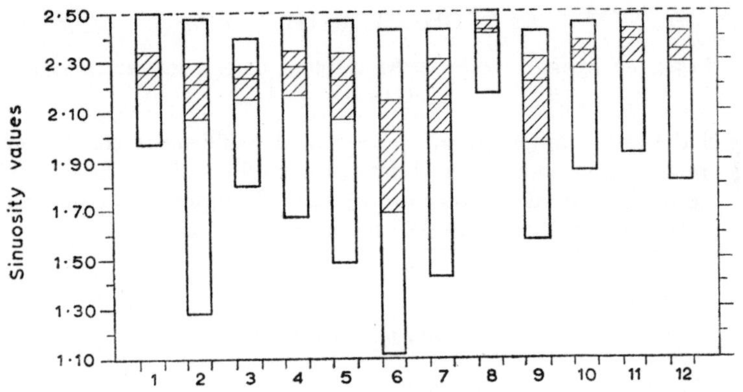

INDEX TO CANALS

1 Huddersfield
2 Leeds & Liverpool
3 Rochdale
4 Trent & Mersey
5 Grand Junction
6 Oxford (Original)
7 Oxford (Improved)
8 Birmingham & Liverpool Junc.
9 Staffs. & Worcs.
10 Worcs. & Birmingham
11 Kennet & Avon
12 Thames & Severn (including Stroudwater)

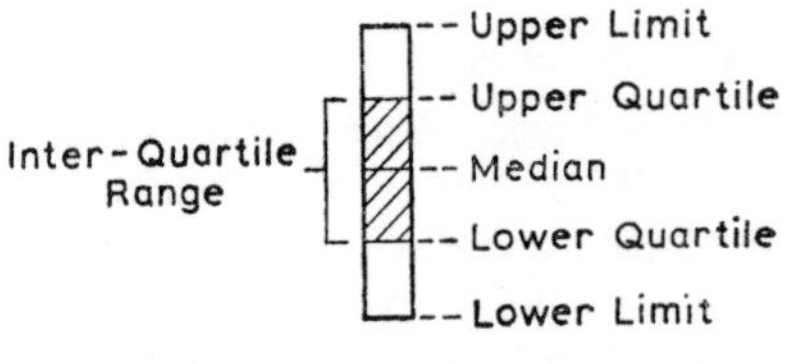

FIG. 22. SINUOSITY VALUE DISTRIBUTION

former exhibits a very strong tendency to straightness; over 75% of its values exceed 2·40. The Oxford Canal shows only a weak tendency to straightness; over 75% of its values are below 2·20, and almost 50% are below 2·00. This high degree of sinuosity is due to the desire on the part of the engineers to avoid all but the most essential earth works, as has been seen (p. 36). The situation is therefore not fundamentally dissimilar from that of other canal lines, and the variation in engineering approach produced an exception of degree rather than of type.

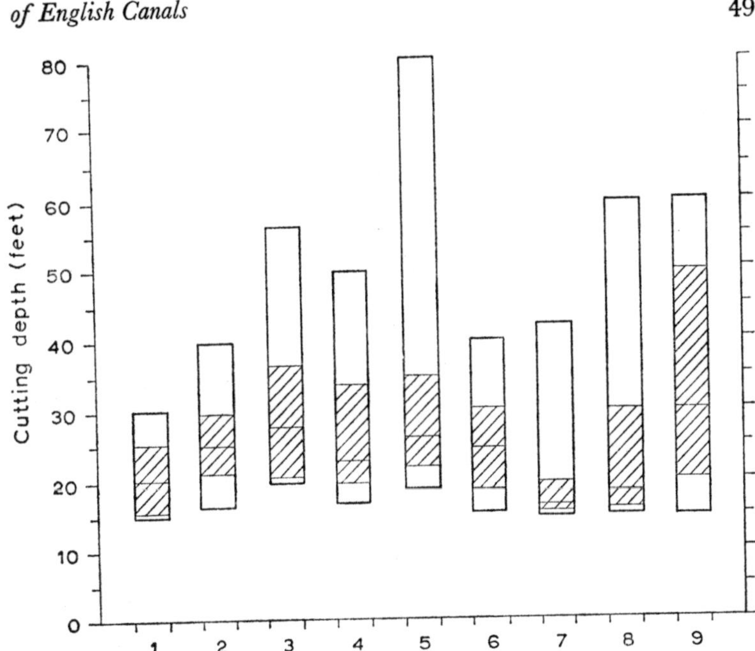

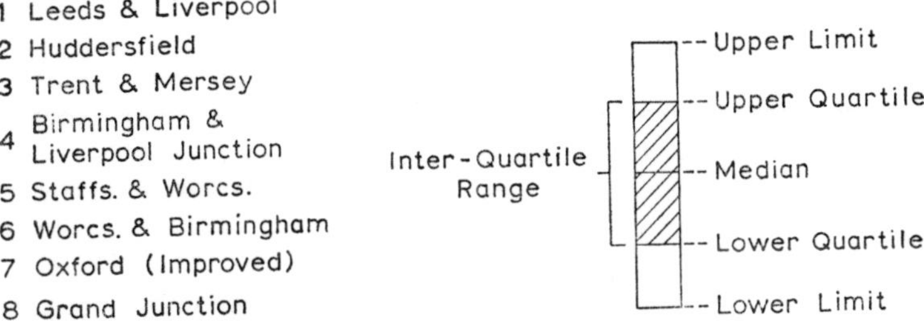

FIG. 23. CUTTING DEPTH DISTRIBUTION

The Birmingham & Liverpool Junction Canal is exceptional in its straightness and relative independence of physiographic controls. The reason for this is important in the context of the development of construction techniques which were to be used widely in railway building, and may be considered separately.

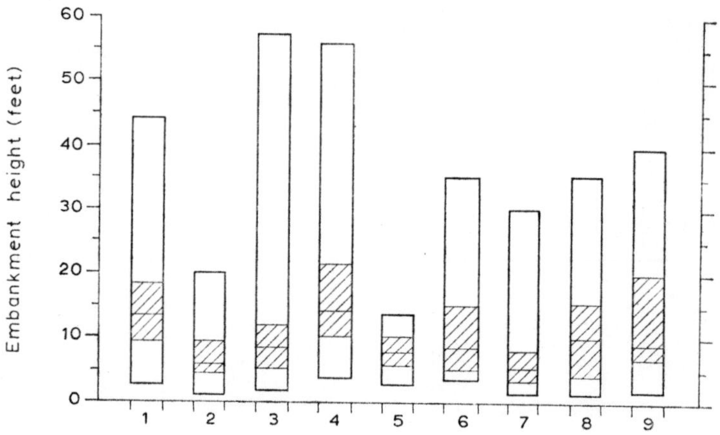

INDEX TO CANALS

1 Leeds & Liverpool
2 Huddersfield
3 Trent & Mersey
4 Birmingham & Liverpool Junc.
5 Staffs. & Worcs.
6 Worcs. & Birmingham
7 Oxford (Improved)
8 Grand Junction
9 Kennet & Avon

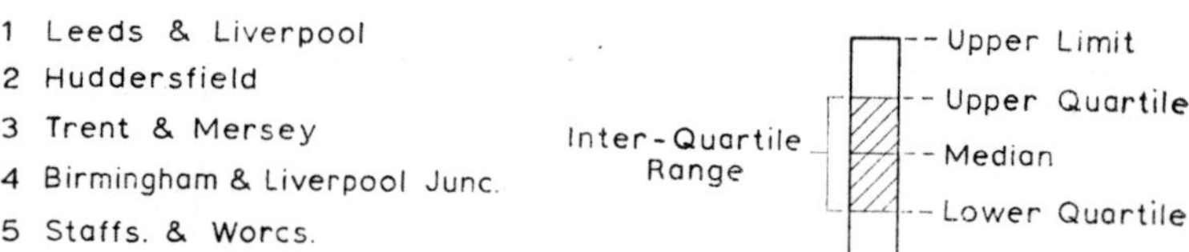

FIG. 24. EMBANKMENT HEIGHT DISTRIBUTION

The Birmingham & Liverpool Junction Canal—engineering concepts

It will be observed from Figures 23 and 24 that the Birmingham & Liverpool Junction Canal does not stand out strongly as having exceptionally deep cutting or high embankment. Thus the popular characterisation of this canal simply as one with exceptionally deep cutting and high embankment is not reflected in these data. There might appear to be a possibility that the uniqueness of the cutting and embankment on this canal lies in the frequency with which it occurs. This is not supported by Figure 25.

What does appear, however, is that this canal has an unusually high proportion of its cutting and embankment as 'true' cutting and embankment; that is, cutting (or embankment) on both sides of the

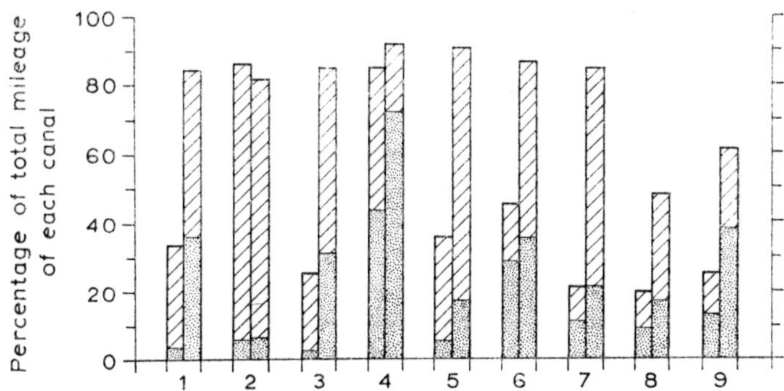

INDEX TO CANALS

1 Leeds & Liverpool
2 Huddersfield
3 Trent & Mersey
4 Birmingham & Liverpool Jn.
5 Staffs. & Worcs.
6 Worcs. & Birmingham
7 Oxford (Improved)
8 Grand Junction
9 Kennet & Avon

Percentage of miles with cutting

Percentage of miles with 'true' cutting

Percentage of miles with embankment

Percentage of miles with 'true' embankment

FIG. 25. PROPORTIONS OF 'TRUE' CUTTING AND EMBANKMENT

canal simultaneously. This is in contrast to the 'contouring' situation common on all earlier canals (Plate 3).

This then explains the fact that the earthworks on this canal enabled its line to be largely liberated from limits imposed by landforms, and enabled it to have a very low degree of sinuosity. The type of earthwork which would perform that function is 'true' cutting and embankment rather than 'contouring' cutting and embankment.

Thus the difference between the engineering concepts behind the Birmingham & Liverpool Junction Canal and behind earlier canals is the high proportion of 'true' cutting and embankment.

'True' cuttings and engineering technique

A less profound variation between earlier and later canals may be noted in Figure 25. The proportion of the number of miles with 'true' cuttings to the total number of miles with cuttings is noticeably higher on later canals. A similar tendency, though less clearly differentiated, may be noted in the case of embankments. This variation between earlier and later canals probably reflects progress made in engineering techniques over the period 1766 to 1826. (See Appendix 2 for dates of Canal Acts). The excavation of 'true' cuttings and the building of 'true' embankments involved moving large amounts of material *along* the canal, in attempts to balance the quantities of cutting and embankment materials. Earlier 'contouring' canals involved the movement of materials *across* the canal (Fig. 20).

STATISTICAL COMPARISONS

Further progress in the analysis of morphological data requires the use of statistical analysis. The Mann-Whitney U Test may be used to determine whether two independent sets of data have been drawn from the same statistical population (Siegel). In other words, it is possible to compare each canal with the other canals and each section (identified on the basis of an assessment of the probable effect of landform differences on canal morphology) with the other sections, using the parameters of sinuosity, cutting depth and embankment height. Analysis for both significant difference and significant similarity can be made with this test, but there is a limitation with respect to the test for significant similarity. Using this procedure, samples should consist of at least 20 observations, due to the statistical nature of the test. Therefore, although it is possible to include all the individual canals in this part of the anlaysis as entities, not all the *sections* may be included, since many are less than 20 miles in length. Indeed, in the case of cutting depths a large proportion of the sections have less than 10 miles in which cutting is recorded, and hence less than 10 values. For this reason it is not practicable to analyse cutting depth by sections.

Before proceeding with the analysis, some points should be made regarding the conceptual framework within which it is located, and of limitations inherent in further analysis. Quantitative measures of morphology have been introduced, and the characteristics of canals have been considered in relation to a wide range of variables including physiographic, economic and engineering factors. Generalisations have been made which have a wide degree of applicability while maintaining depth of detail.

The further analysis which follows is an attempt to obtain greater precision by making objective comparisons between the morphology of each canal or section. But as the analysis becomes more selective (in terms of number of variables) and more rigid (in terms of the analytical framework of canals and sections), the applicability of generalisations tends to decrease. The conditions which have to be fulfilled before a relationship or 'linkage' is accepted as significant must become more stringent in the desire for greater precision and the tendency ultimately is for there to be so few linkages that generalisations cannot be made.

Therefore although useful additions are made to the conclusions already presented, and although a greater element of precision is gained by these additions, the conclusions already made remain valid. Indeed, they are in many cases confirmed by the results of further analysis. Two types of result can be obtained from the application of the analysis:

(*a*) A result may be obtained from the test for significant *difference* by using the number of linkages which each canal or section of canal possesses; that is, the number of canals (or sections) from which each canal (or section) *differs* significantly. In this way an indication is given of the degree to which a certain canal or section is 'exceptionally different'. Such indications may not be taken too far in formulating generalisations, but are useful in confirming points already made.

(*b*) The test for significant *similarity* identifies canals and sections which are significantly *similar* to each other; the reasons for these similarities may then be discussed.

The analysis is applicable to five sets of data:

(i) Sinuosity values (whole canals)
(ii) Cutting depths (whole canals)
(iii) Embankment heights (whole canals)
(iv) Sinuosity values (canal sections)
(v) Embankment heights (canal sections)

The results of the analysis are discussed below; first the test for significant difference, and second the test for significant similarity.

Significant Difference

Figures 26–30 show in graph form the number of bonds or linkages of significant difference for each canal or section in each of the five sets of data. The level of significance is taken at 0·2%

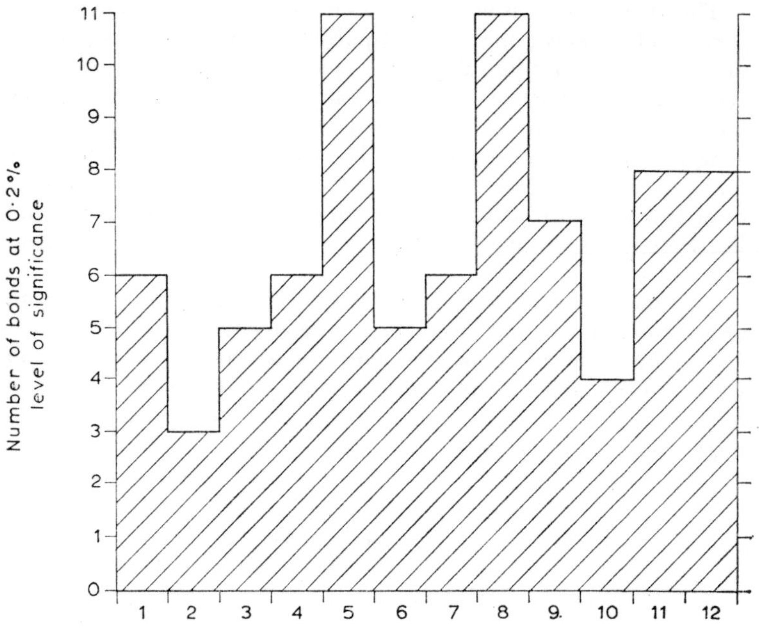

INDEX TO CANALS

1 Leeds & Liverpool
2 Huddersfield
3 Rochdale
4 Trent & Mersey
5 Birmingham & Liverpool Junc.
6 Staffs. & Worcs.
7 Worcs. & Birmingham
8 Oxford (Original)
9 Oxford (Improved)
10 Grand Junction
11 Kennet & Avon
12 Thames & Severn (including Stroudwater)

FIG. 26. BONDS OF SIGNIFICANT DIFFERENCE—SINUOSITY (whole canals)

probability; the chances that these differences could occur by chance are 2 in 1,000.

(i) *Sinuosity values* (*whole canals*). Consider first the case of sinuosity values for whole canals (Fig. 26). The graph showing the number of bonds illustrates the exceptional degree to which both the Oxford (with *original* northern line) and Birmingham & Liverpool Junction Canals are significantly different; each differs from all the other canals, and from each other. This bears out the points already made

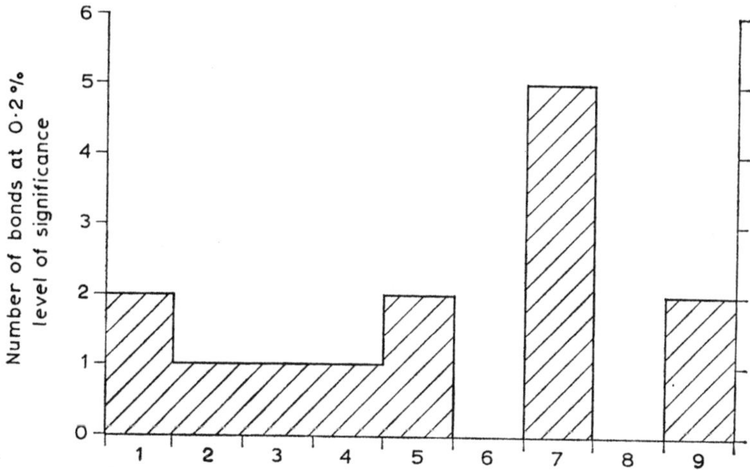

INDEX TO CANALS

1 Leeds & Liverpool
2 Huddersfield
3 Trent & Mersey
4 Birmingham & Liverpool Junc.
5 Staffs. & Worcs.
6 Worcs. & Birmingham
7 Oxford (improved)
8 Grand Junction
9 Kennet & Avon

Fig. 27. Bonds of Significant Difference—Cutting Depth (whole canals)

about the exceptional nature of these two extreme forms of canal construction (p. 47).

(ii) *Cutting Depths* (*whole canals*). Next consider cutting depths for each canal (Fig. 27). The Oxford Canal (with *improved* northern line) is seen to be strongly different from five of the other canals, while none of the others has more than two bonds. This is almost certainly due to the exceptionally small amount of cutting on the Oxford Canal, even taking into account the improved northern section with its cuttings. (Cutting data is not available for the original northern line). It is important to note that the Birmingham & Liverpool Junction Canal does not stand out as strongly different from other canals, having only one bond at this level of significance. This is in line with the earlier conclusion, that it is the type of cutting rather than its depth which distinguishes the Birmingham & Liverpool Junction Canal (p. 50).

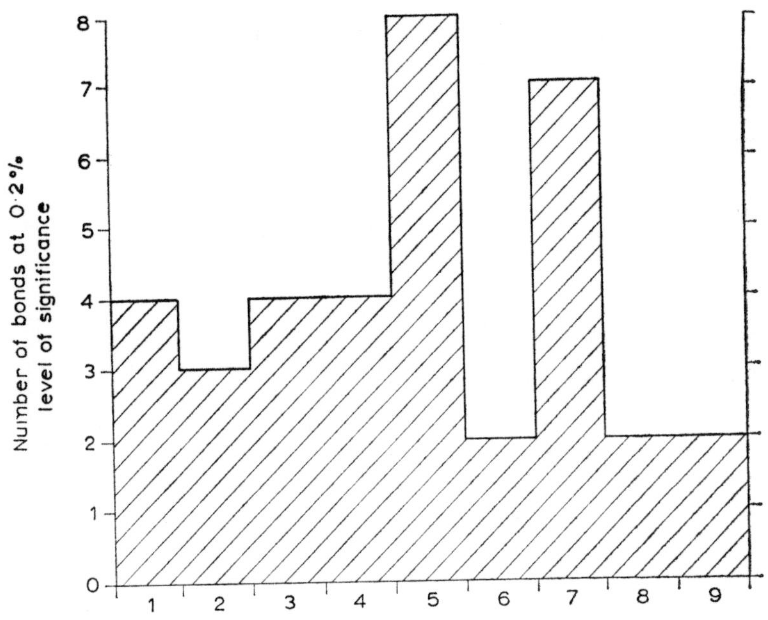

INDEX TO CANALS

1 Leeds & Liverpool
2 Huddersfield
3 Trent & Mersey
4 Birmingham & Liverpool Junc.
5 Staffs. & Worcs.
6 Worcs. & Birmingham
7 Oxford (improved)
8 Grand Junction
9 Kennet & Avon

FIG. 28. BONDS OF SIGNIFICANT DIFFERENCE—EMBANKMENT HEIGHT (whole canals)

(iii) *Embankment Heights* (*whole canals*). The bonds occurring in the matrix of embankment data (Fig. 28) echo the results of the analysis of cutting data, in that the Oxford Canal is again shown to be strongly exceptional. The Staffordshire & Worcestershire Canal also exhibits a large number of bonds. This is probably due to the frequent occurrences of deep cuttings on the Stour–Smestow section of the canal (Fig. 16). Again, on the basis of embankment height, the Birmingham & Liverpool Junction Canal does not appear to be exceptional (p. 50).

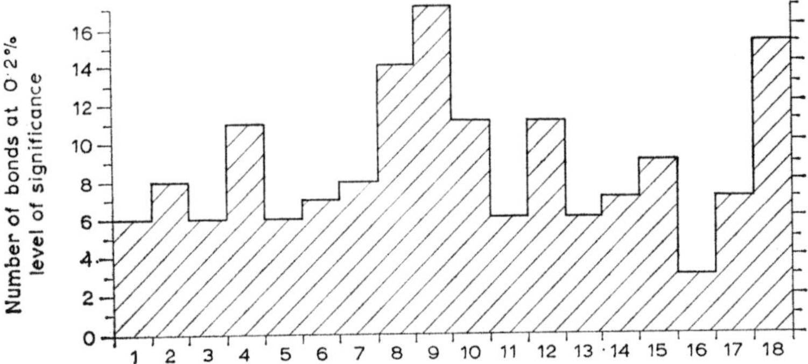

INDEX TO CANAL SECTIONS

1	Rochdale	0-16 (miles)	10	Oxford	54-72
2	Leeds & Liverpool	0-27	11	Oxford (Improved N.line)	0-24
3	Leeds & Liverpool	37-52	12	Oxford	54-72
4	Leeds & Liverpool	52-83	13	Grand Junction	5-20
5	Leeds & Liverpool	94-117	14	Grand Junction	20-53
6	Trent & Mersey	0-20	15	Grand Junction	58-76
7	Trent & Mersey	48-67	16	Staffs. & Worcs.	29-47
8	Trent & Mersey	75-98	17	Worcs. & Birmingham	0-17
9	Oxford (Original N.line)	0-31	18	Kennet & Avon	48-76

FIG. 29. BONDS OF SIGNIFICANT DIFFERENCE—SINUOSITY (sections)

(iv) *Sinuosity values* (*sections*). The graph of bonds between sections, using the sinuosity variable, shows that three sections are 'exceptionally different': Miles 75–98 of the Trent & Mersey Canal; miles 0–31 of the original northern line of the Oxford Canal; and miles 48–76 of the Kennet & Avon Canal (Fig. 29). The ways in which these sections differ from the rest of the sections can be deduced from a knowledge of the characteristics of each section. The section of the Trent & Mersey between miles 75 and 98 has an exceptionally low degree of sinuosity, made possible by the ease of construction of the canal in the Trent valley below Alrewas; this has been discussed (p. 45). The section of the Kennet & Avon between miles 48 and 76 has similar characteristics and also passes along an open well-developed river valley—that of the Kennet (p. 45). The section of the original line of the Oxford Canal between miles 0 and 31 represents the opposite extreme, being highly sinuous in its 'contouring' traverse of the head of the Avon basin (p. 36).

(v) *Embankment Heights* (*sections*). Finally, analysis of the sections on the basis of the embankment height variable reveals three ex-

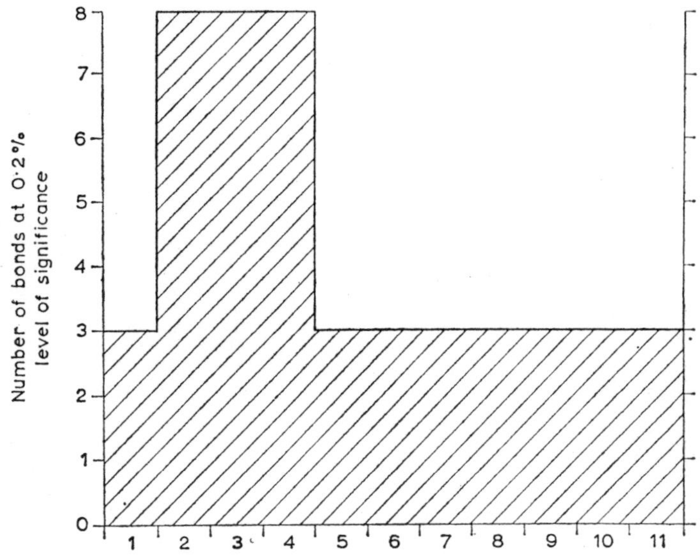

INDEX TO CANAL SECTIONS

1	Leeds & Liverpool	0-27 (miles)	7	Oxford (Improved N. line)	0-24
2	Leeds & Liverpool	52-83	8	Oxford	54-72
3	Leeds & Liverpool	94-117	9	Staffs. & Worcs.	29-47
4	Trent & Mersey	0-20	10	Worcs. & Birmingham	0-17
5	Trent & Mersey	48-67	11	Kennet & Avon	48-76
6	Trent & Mersey	75-98			

FIG. 30. BONDS OF SIGNIFICANT DIFFERENCE—EMBANKMENT HEIGHT (sections)

ceptional sections: miles 52–83 of the Leeds & Liverpool; miles 94–117 of the Leeds & Liverpool; and miles 0–20 of the Trent & Mersey (Fig. 30). The first section has an exceptionally large number of high embankments in traversing the dissected slopes of the North Rossendale scarp between Feniscowles, 5 miles west of Blackburn, and Foulridge tunnel. The second section does not show any exceptional features in its embankment height profile, apart from a rather more consistent tendency for moderately high embankment (10–20 ft.) to occur throughout its length. This may be the reason for its appearance as an 'exceptional' section, but this explanation must remain tentative. The third section, like the first, is very heavily embanked in its negotiation of the steep dissected side of the Weaver valley (p. 41).

Therefore the use of the test for significant difference serves to reinforce independently some of the conclusions made in earlier chapters.

Significant similarity

It remains to discuss the results of the second part of the analysis which is the identification of bonds of significant similarity between canals, and between canal sections. The selection of a level of significance presents some difficulty. It would obviously be desirable to maintain the same level of significance as was used in the analysis of significant difference—that is a 0·2% probability level. However, the accompanying table (Table 2) shows that the highest level of significance at which a bond of similarity occurs is 0·80% probability.

It is therefore not advisable to use these results as the sole basis for generalisations. However, the more significant associations and their implications may be considered, and serve to reinforce previously stated findings.

Similarity and landforms

The section of the Leeds & Liverpool between miles 52 and 83 is bonded with the section of the Oxford Canal between miles 54 and 72 on the basis of sinuosity at the level of 0·80%. The Leeds & Liverpool section traverses the strongly dissected slope of the North Rossendale Uplands, and the Oxford section follows the mature trough-like meandering valley of the Cherwell.

The Leeds & Liverpool section 0–27 is bonded with the Kennet & Avon section 48–76 on the basis of embankment height at the level of 3·97%. Again the two bonded sections cross landforms of different genetic type. The Leeds & Liverpool section traverses the low, gentle slope which bounds the eastern edge of the West Lancashire Plain. The Kennet & Avon section follows the gently sloping side of the Kennet valley, or is situated in the river flood plain. Both situations require the use of only low embankment, generally on the towing path side of the canal alone.

Finally, the Trent & Mersey section 48–67 is bonded with the Oxford section 0–24 (improved northern line) on the basis of embankment height, at the level of 2·40%. Yet again a section traversing a scarp slope is associated with a section following a river valley. The Trent & Mersey section follows the Trent valley, which has moderately well-dissected sides, although the canal is below the level at which these dissections enter the main valley.

TABLE 2—BONDS OF SIGNIFICANT SIMILARITY WITH A HIGHER LEVEL OF SIGNIFICANCE THAN 20%

Variable	Canals or Sections Associated				% Level of Significance
Sinuosity	L & L		Staffs. & Worcs.		1·60
(Whole Canals)	Rochdale		Gr. Junc.		11·14
Cutting	L & L		Gr. Junc.		9·56
(Whole Canals)	T & M		K & A		3·98
Embankment	T & M		Gr. Junc.		3·98
(Whole Canals)					
Sinuosity	L & L	52–83*	Ox.	54–72	0·80
(Canal Sections)	Rochdale	0–16	L & L	37–52	12·72
	L & L	0–27	Gr. Junc.	58–76	11·14
	T & M	48–67	Gr. Junc.	58–76	15·86
	L & L	0–27	T & M	48–67	18·20
Embankment	L & L	0–27	K & A	48–76	3·97
(Canal Sections)	T & M	48–67	Ox**	0–24	2·40
	T & M	75–98	W & B	0–17	11·92

L & L — Leeds & Liverpool Canal
T & M — Trent & Mersey Canal
K & A — Kennet & Avon Canal
W & B — Worcester & Birmingham Canal

* Figures refer to mileages
** Improved northern line

The river meanders impinge in a number of places on the line of the canal, and the section includes a river crossing by aqueduct; these features involve a number of moderately high embankments (5 to 15 ft.). The Oxford Canal section traverses gentle slopes around the head of the Stratford Avon valley, but higher embankment occurs on the 1829 improvements (up to 15 ft.).

These associations between sections which traverse landforms of different genetic types suggests that similar physical features (the significant factors being slope gradient and slope sinuosity) may be presented for canal construction by landforms of different genetic types. It has already been shown that similar physical features may be presented for canal construction by landforms of similar genetic types (pp. 45 and 57).

Similarity and date of construction

The Leeds & Liverpool Canal is associated with the Staffordshire & Worcestershire Canal on the basis of sinuosity at the 1·60% probability level. It is interesting to note that both canals were originally laid out by James Brindley and his pupils in the early stages of the canal era, though a considerable portion of the Leeds & Liverpool Canal's line, between Foulridge and Newburgh, was later changed. Even then, however, Robert Whitworth, Brindley's pupil and son-in-law, was responsible for most of its length. No doubt Whitworth's engineering outlook was much influenced by Brindley's strong opinions.

The Trent & Mersey is associated with the Kennet & Avon on the basis of cutting depth at the level of 3·98%. This is an interesting association, lying as it does across differences of engineer and date (see Appendix 2). The Trent & Mersey is associated with the Grand Junction on the basis of embankment height at the level of 3·98%. Again the Trent & Mersey is associated with a later canal, laid out by a different engineer. These significant similarities, lying across differences of engineer and date, reinforce the earlier conclusion (p. 47) that similar factors operated throughout the span of time and space covered by the construction of the canals, due to basically uniform principles of construction (the Birmingham & Liverpool Junction Canal excepted).

The Hypotheses

The two general hypotheses (p. 25) to be tested were:

(*a*) Canal morphology (as measured by the indices) varies with landform;

(*b*) canal morphology varies with the advance in engineering knowledge during the period of canal construction.

It has been shown that (*a*) is borne out everywhere on the canals sampled, and it is reasonable to say that this will apply to the whole network. Hypothesis (*b*) is also borne out, though it is clear that basically similar constructional methods were practised during most of the period of canal construction in England. It was only in the early 19th century that sufficient advances in technique were made for a significant variation in morphology to be made. This is indicated by the Birmingham & Liverpool Junction Canal (p. 51). The argument may be taken a step further, and the two hypotheses linked, in the following way. Given similar construction methods through almost all of the period of canal construction in England, it is reasonable to point to factors of landform as being the most important in determining the variations which have been observed in canal morphology.

PLATE 6 CAEN HILL LOCKS, KENNET & AVON CANAL
Devizes is visible beyond the locks. Each short pound between locks has its side-pond, to the left in the view. Since this photograph was taken, the water level has fallen considerably, and the side-ponds are now used as allotments and paddocks. Note the wide locks—one cause of water shortages.
Looking east.
(Aerofilms Ltd.)

Conclusion

The factors which influenced the routes of canals and their morphology were very varied. Evidence for the operation of these factors is also varied, and falls into two main categories—contemporary and 'physical'. The former consists mainly of archive material, and limitations of space have prevented detailed consideration of this source. The latter group of evidence consists of morphological measurement, and map and field observations. By combining the approaches based on these sources in a study which focuses on canal morphology, a number of relationships between canal and 'environment' may be observed. In order to put these conclusions in the perspective of all factors influencing canal morphology, the relevant factors may be summarised as follows:

Economic factors determined the broad route of the canal—the termini and the important industrial or agricultural areas to be served between the termini.

Physiographic factors, acting through engineering ability, moulded the route into a practicable line with a specific morphology. The degree to which a particular route was deviated by these factors depended on the cost of overcoming physical obstacles and on the engineering ability to do so. The engineering principles embodied in the Birmingham & Liverpool Junction Canal mark the stage at which physiographic influence was very much reduced.

The effect of vested interests was usually to produce only minor alterations in the optimum compromise physiographic/economic line. These alterations might be small deviations, perhaps with relatively heavy cutting and embankment, or the designing of bridges and tunnels to blend with landscaped estates; or the arrangement of the towing path on the 'inside' bank of a slope-traversing canal. The Leeds & Liverpool Canal is an exception in that vested interests played an important part in a major change of route (Farrington, 1970).

Given a basic desire for a line to connect terminal areas, and similar construction methods for most of the period in question, physiographic factors were the most significant parts of 'total environment' in affecting canal morphology, both large-scale and small-scale, for almost all of the period of canal construction in England.

ABSTRACT

Morphological Studies of English Canals

In a previous paper in this series (*A Morphological Approach to the Geography of Transport*, Occasional Papers in Geography No. 3, University of Hull, 1965), J. H. Appleton explained the need in transport studies for the morphological approach as a complement to the functional approach. The present paper adopts this approach to the English trunk canal system built up in the late 18th and early 19th centuries. It examines the relationships between canal morphology and 'external' variables such as landforms, water supply, constructional problems, and canal promoters' objectives. The measurement of morphological features—sinuosity, cutting depth and embankment height—and the testing of statistical significance is introduced to aid in generalising about these relationships.

The basic conclusion reached is that given similar construction methods throughout most of the period involved, physiographic factors were the most important in determining canal morphology during this time.

Economic factors determined the broad route of the canals—the termini and the significant agricultural or industrial areas to be served between the termini.

Physiographic factors, acting through the medium of engineering ability, moulded the route into a practicable line with a specific morphology.

The effect of vested interests (notably other canal companies) was usually to produce only minor alteration in the optimum line of the canal.

RÉSUMÉ

ÉTUDES MORPHOLOGIQUE DES CANAUX ANGLAIS

Dans un article précédent de cette même série (*A Morphological Approach to the Geography of Transport*), J. H. Appleton a expliqué la nécessité d'études de transport proprement dites pour l'approche morphologique comme complément à l'approche fonctionnelle du transport. On se sert, dans le présent exposé, de cette approche pour étudier le système anglais de canaux principaux, construits à la fin du XVIIIe et au début du XIXe siècle. On y étudie les relations entre la morphologie des canaux et les variantes "externes", telles que les formes du terrain, l'approvisionnement en eau, les problèmes de construction et les objectifs des fondateurs de ces canaux. On y introduit la mesure des traits morphologiques—sinuosité, profondeur de creusement, et hauteur des berges—ainsi que l'examen des significations statistiques afin de parvenir à la généralisation de ces relations.

La conclusion de base à laquelle on est parvenu est que, étant données les méthodes de construction semblables tout au long de la période considérée, c'étaient les facteurs physiographiques qui avaient la plus grande importance sur la détermination de la morphologie des canaux à cette époque.

Des facteurs économiques déterminaient le tracé général des canaux—les termini et les régions agricoles et les zônes industrielles qui devaient être desservies entre ces termini.

Les facteurs physiographiques, opérant au travers des possibilités techniques du génie modelèrent le tracé en une ligne praticable ayant une morphologie spécifique.

L'incidence des intérêts particuliers (notament les autres compagnies de canaux) ne produisaient généralement que des altérations mineures quant au tracé le plus satisfaisant du canal.

ABRISS

Morphologische Untersuchungen der englischen Kanäle

In einem früheren Artikel in dieser Serie (*A Morphological Approach to the Geography of Transport*) erklärte J. H. Appleton die Notwendigkeit einer morphologischen Betrachtungsweise in Verkehrsstudien, zusätzlich zu der funktionalen. Dieser Artikel wendet diese Methode auf das englische Hauptkanalsystem an, das im späten 18. u. frühen 19. Jahrhundert ausgebaut wurde. Er untersucht das Verhältnis zwischen Kanalmorphologie und 'äusseren' Variablen wie Bodenformationen, Wasserzufuhr, Konstruktionsproblemen und den Zielen und Zwecken der Kanalauftraggeber. Die Messwerte morphologischer Charakteristika—wie Windung, Einschnitt–Tiefe und Höhe der Böschung—und die Untersuchung der statistischen Bedeutung werden angeführt, um bei der Herausarbeitung allgemeiner Prinzipien über diese Beziehungen zu helfen.

Es ergab sich die grundsätzliche Schlussfolgerung, dass bei ähnlichen Konstruktionsmethoden, während des Grossteils dieser Epoche, die physiographischen Faktoren während dieser Zeit den Ausschlag über die Kanalmorphologie gaben.

Wirtschaftliche Faktoren bestimmten den ungefähren Verlauf des Kanals—die Anfangs-und Schlusspunkte und die bedeutenden landwirtschaftlichen oder industriellen Gebiete entlang des Kanals, die dadurch versorgt wurden.

Physiographische Faktoren und die damaligen technischen Möglichkeiten bestimmten den tätsachlichen Verlauf des Kanals und seine spezifische Morphologie.

Andere rechtmässige Interessen (v.a. die anderer Kanalgesellschaften) bewirkten gewöhnlich nur geringe Abweichungen vom günstigsten Verlauf des Kanals.

References

J. H. Appleton, 1962. *The Geography of Communications in Great Britain*, p. 7 1963. The Efficacy of the Great Australian Divide as a Barrier to Railway Communication, *Transactions of the Institute of British Geographers*, No. 33, p. 107. 1965. A Morphological Approach to the Geography of Transport, *University of Hull Occasional Papers in Geography*, No. 3.

W. W. Bishop, 1958. The Pleistocene Geology and Geomorphology of Three Gaps in the Midland Jurassic Scarp, *Philosophical Trans. Royal Soc.*, Vol. 241, p. 255.

B.T.H.R., KAC 3/12. Whitworth's report to the Kennet & Avon Committee, 25th July, 1789.
KAC 4/12, p. 25. Smith's report, 1812.

J. H. Farrington, 1969. *Morphological Studies of English Canals*, unpublished Ph.D. Thesis, University of Hull. 1970. The Leeds & Liverpool Canal—a study in route selection, *Transport History*, Vol. 3, No. 1, p. 52.

C. Hadfield, 1966. *The Canals of the West Midlands.*

D. L. Linton, 1932. The Origin of Wessex Rivers, *Scottish Geographical Magazine*, Vol. 48, p. 149.

F. W. Shotton, 1953. Pleistocene Deposits of the area between Coventry, Rugby and Leamington and their bearing upon the Topographic Development of the Midlands, *Philosophical Trans. Royal Soc., Series B*, Vol. 237, p. 209.

S. Siegel, 1956. *Nonparametric Statistics for the Behavioural Sciences*, p. 116.

L. J. Wills, 1938. The Pleistocene Development of the Severn from Bridgnorth to the Sea, *Quart. J. of the Geol. Soc. of London, Vol. 94, p. 226.*

Canal Acts

Birmingham Canal, 1768, 8 Geo III c.38.

Birmingham & Liverpool Junction Canal, 1826, 7 Geo IV c.95.

Coventry Canal, 1768, 8 Geo III c.36.

Grand Junction Canal, 1793, 33 Geo III c.80.

Grand Union Canal, 1810, 50 Geo III c.122.

Huddersfield Canal, 1794, 34 Geo III c.53.

Kennet & Avon Canal, 1794, 34 Geo III c.90.

Leeds & Liverpool Canal, 1770, 10 Geo III c.114.

Leicestershire & Northamptonshire Union Canal, 1793, 33 Geo III c.98.

Oxford Canal, 1769, 9 Geo III c.70.

Oxford Canal (improved northern line), 1829, 10 Geo IV c.48.

Rochdale Canal, 1794, 34 Geo III c.78.

Somersetshire Coal Canal, 1794, 34 Geo III c.86.

Staffordshire & Worcestershire Canal, 1766, 6 Geo III c.97.

Thames & Severn Canal, 1783, 23 Geo III c.38.

Trent & Mersey (Grand Trunk) Canal, 1766, 6 Geo III c.96.

Warwick & Birmingham Canal, 1793, 33 Geo III c.38.

Warwick & Napton Canal (originally Warwick & Braunston, 1794) 1796, 36 Geo III c.95.

Wiltshire & Berkshire Canal, 1795, 35 Geo III c.52.

Worcester & Birmingham Canal, 1791, 31 Geo III c.59.

Wyrley & Essington Canal, 1792, 32 Geo III c.81.

APPENDIX I

CANALS SELECTED FOR STUDY IN ORIGINAL RESEARCH

1. Leeds & Liverpool Canal.
2. Rochdale Canal. [a] [b]
3. Huddersfield Canal. [b]
4. Trent & Mersey Canal ('The Grand Trunk').
5. Birmingham & Liverpool Junction Canal (now part of the Shropshire Union system).
6. Staffordshire & Worcestershire Canal.
7. Worcester & Birmingham Canal.
8. Oxford Canal.
9. Grand Junction Canal (now part of the Grand Union system).
10. Kennet & Avon Canal. [b]
11. Thames & Severn Canal [a] [b] (with the Stroudwater Canal). [a] [b]

a Waterways not administered by the British Waterways Board (in 1972.)
b Disused for navigation for all or part of length (in 1972).

Appendix 2

ENGINEERS RESPONSIBLE FOR CANAL LINES, AND DATES OF ACTS

Canal	Date of Act under which work was begun	Engineer(s) responsible for original line	Line opened
Leeds & Liverpool	1770	John Longbotham Josiah Clowes James Brindley Robert Whitworth	1816
Rochdale	1794	John Rennie	1804
Huddersfield	1794	Benjamin Outram Josiah Clowes	1811
Trent & Mersey	1766	John Smeaton James Brindley	1777
Birmingham & Liverpool Junction	1826	Thomas Telford	1835
Staffordshire & Worcestershire	1766	James Brindley	1771
Worcester & Birmingham	1791	John Snape Josiah Clowes Thomas Cartwright	1815
Oxford	1769	James Brindley Samuel Simcock	1790
Oxford (improved northern line)	1829	Charles Vignoles	1834
Grand Junction	1793	James Barnes William Jessop	1800[1]
Kennet & Avon	1794	Samuel Weston Robert Whitworth John Rennie	1810
Thames & Severn	1783	James Brindley Robert Whitworth	1789

[1]A tramway operated over Blisworth Hill until Blisworth Tunnel was completed in 1805.